Vitamin D in Health and Disease (2nd Edition)

Vitamin D in Health and Disease (2nd Edition)

Guest Editors

Giuseppe Murdaca
Sebastiano Gangemi

MDPI

Basel • Beijing • Wuhan • Barcelona • Belgrade • Novi Sad • Cluj • Manchester

Guest Editors
Giuseppe Murdaca
Department of Internal Medicine
University of Genoa and San Bartolomeo Hospital
Sarzana
Italy

Sebastiano Gangemi
Department of Clinical and Experimental Medicine
University of Messina
Messina
Italy

Editorial Office
MDPI AG
Grosspeteranlage 5
4052 Basel, Switzerland

This is a reprint of the Special Issue, published open access by the journal *Biomedicines* (ISSN 2227-9059), freely accessible at: https://www.mdpi.com/journal/biomedicines/special_issues/vitamin_D_health_dis_2.

For citation purposes, cite each article independently as indicated on the article page online and as indicated below:

Lastname, A.A.; Lastname, B.B. Article Title. *Journal Name* **Year**, *Volume Number*, Page Range.

ISBN 978-3-7258-5675-6 (Hbk)
ISBN 978-3-7258-5676-3 (PDF)
https://doi.org/10.3390/books978-3-7258-5676-3

Contents

biomedicines

MDPI

Editorial

Vitamin D in Health and Disease 2.0

Giuseppe Murdaca [1,*] and Sebastiano Gangemi [2]

[1] Allergology and Clinical Immunology Unit, Department of Internal Medicine, San Bartolomeo Hospital, University of Genoa, 19038 Sarzana, Italy
[2] Allergy and Clinical Immunology Unit, Department of Clinical and Experimental Medicine, University of Messina, Via Consolare Valeria, 98125 Messina, Italy; gangemis@unime.it
* Correspondence: giuseppe.murdaca@unige.it; Tel.: +39-0103537924

Vitamin D (VD) is a fat-soluble vitamin considered essential for human health, and its levels are associated with the function and composition of the intestinal microbiome. In fact, it plays a decisive role in the pathogenetic process of immunological diseases related to the intestinal microbiome starting from the pediatric age; its integration is in fact capable of increasing the diversity of the microbiome in children, with a consequent positive impact on the functionality of the immune system. It has been demonstrated that VD is capable of increasing the number and variety of commensal bacteria, which are able to compete with pathogenic bacteria for nutrients, thus exercising a synergistic protective function against various diseases [1].

Recent evidence has also highlighted the potential role of VD in allergic and immune-mediated diseases, exerted by its ability to modify the immune response and the proliferation and function of mast cells, particularly through the modulation of the IL-31/IL-33 genes axis [2]. Increasing its concentration within the intestinal bacteria responsible for the release of short-chain fatty acids would increase the secretion of immunoglobin A by B cells, reducing macrophages and dendritic cells and limiting chemotaxis in the neutrophils. All this would confirm the fundamental role of VD in innate and adaptive immunity, in addition to its anti-inflammatory properties.

Free serum VD is stabilized and transported to target tissues by VD-binding proteins (VDBPs), where VD exerts its biological effect at the cellular level by binding to the VD ligand-activated receptor (VDR), belonging to the superfamily of nuclear receptors. After binding to the biologically active form of VD, VDR enters the nucleus and forms a heterodimer with the retinoid X receptor, thus regulating gene transcription. Several subtle allelic polymorphisms have been identified in the gene encoding VDR and have been linked to diseases such as osteoporosis and metabolic bone disorders, although their correlation has not yet been fully understood. VDR is also implicated in the renal and intestinal absorption of calcium and phosphate, as well as the downregulation of PTH synthesis, and it appears to play a role in the management of persistent secondary hyperparathyroidism after renal transplantation.

Beyond the role of VD transporter, VDBP also has immunological and anti-inflammatory properties. This happens, for example, in the context of multiple sclerosis, where the variation in VDBP levels, in synergy with VD insufficiency, would negatively influence the severity and duration of the disease [3].

VD is also involved in fighting infectious processes, where it promotes the synthesis of antimicrobial peptides, suggesting a likely mechanism by which this molecule could provide protection against some respiratory pathogens, including severe acute respiratory syndrome coronavirus-2, although current evidence is inconsistent regarding its effect on the outcomes and length of hospital stay of such patients [4]. Notably, VD deficiency favors the chronicity of hepatitis B virus infection and, consequently, the degree of fibrosis and the progression to liver cirrhosis.

Finally, VD plays a fundamental and widely demonstrated role in the prevention of oxidative stress. Indeed, its deficiency induces the interruption of mitochondrial function and is associated with markers of inflammation. Furthermore, the conversion of 25(OH)D3 into 1,25(OH)2D3 occurs at the extrarenal tissue level by stimulating the production of macrophages through the type 2 IFN response. This condition is particularly expressed in the context of human exposure to organochlorine pesticides, potentially capable of determining the interruption of the conversion of 25(OH)D3 up to '1,25(OH)2D3. Interference with the function of VDRs in the body and impairment of calcium homeostasis results in VD deficiency and increased risk of developing chronic conditions related to increased oxidative stress and chronic inflammation, such as diabetes and cardiovascular diseases [5]. The same mechanism is also found in the context of polycystic ovary syndrome, where VD integration may be able to reduce the levels of oxidative stress involved in the pathogenetic mechanisms of this pathology in synergy with other chronic pathologies.

Conflicts of Interest: The authors declare no conflicts of interest.

References

1. Tabassum, A.; Ali, A.; Zahedi, F.D.; Ismail, N.A.S. Immunomodulatory Role of Vitamin D on Gut Microbiome in Childre *Biomedicines* **2023**, *11*, 1441. [CrossRef] [PubMed]
2. Murdaca, G.; Allegra, A.; Tonacci, A.; Musolino, C.; Ricciardi, L.; Gangemi, S. Mast Cells and Vitamin D Status: A Clinical an Biological Link in the Onset of Allergy and Bone Diseases. *Biomedicines* **2022**, *10*, 1877. [CrossRef] [PubMed]
3. Hajeer, S.; Nasr, F.; Nabha, S.; Saab, M.B.; Harati, H.; Desoutter, A.; Al Ahmar, E.; Estephan, E. Association between vitamin D deficiency and multiple sclerosis- MRI significance: A scoping review. *Heliyon* **2023**, *9*, e15754. [CrossRef] [PubMed]
4. Renieris, G.; Foutadakis, S.; Andriopoulou, T.; Spanou, V.M.; Droggiti, D.E.; Kafousopoulos, D.; Gkavogianni, T.; Damorak G.; Vatsellas, G.; Giamarellos-Bourboulis, E.J. Association of Vitamin D with severity and outcome of COVID-19: Clinical an Experimental Evidence. *J. Innate Immun.* 2023; *in press*.
5. Yang, J.H.; Lee, Y.M.; Bae, S.G.; Jacobs, D.R., Jr.; Lee, D.H. Associations between organochlorine pesticides and vitamin D deficiency in the U.S. population. *PLoS ONE* **2012**, *7*, e30093. [CrossRef] [PubMed]

biomedicines

MDPI

Review

Mast Cells and Vitamin D Status: A Clinical and Biological Link in the Onset of Allergy and Bone Diseases

Giuseppe Murdaca [1,*,†], Alessandro Allegra [2,*,†], Alessandro Tonacci [3], Caterina Musolino [2], Luisa Ricciardi [4] and Sebastiano Gangemi [4]

1 Department of Internal Medicine, Ospedale Policlinico San Martino, 16132 Genoa, Italy
2 Department of Human Pathology in Adulthood and Childhood "Gaetano Barresi", Division of Hematology, University of Messina, 98125 Messina, Italy; cmusolino@unime.it
3 Clinical Physiology Institute, National Research Council of Italy (IFC-CNR), 56124 Pisa, Italy; atonacci@ifc.cnr.it
4 Department of Clinical and Experimental Medicine, School and Operative Unit of Allergy and Clinical Immunology, University of Messina, 98125 Messina, Italy; luisa.ricciardi@unime.it (L.R.); gangemis@unime.it (S.G.)
* Correspondence: giuseppe.murdaca@unige.it (G.M.); aallegra@unime.it (A.A.)
† These authors have equally contributed.

Abstract: The immune system is made up by an extremely composite group of cells, whose regulated and harmonious activity is fundamental to maintain health. The mast cells are an essential effector of inflammatory response which is characterized by a massive release of mediators accumulated in cytoplasmic secretory granules. However, beyond the effects on immune response, mast cells can modify bone metabolism and are capable of intervening in the genesis of pathologies such as osteoporosis and osteopenia. Vitamin D is recognized to induce changes in bone metabolism, but it is also able to influence immune response, suppressing mast cell activation and IgE synthesis from B cells and increasing the number of dendritic cells and IL-10-generating regulatory T cells. Vitamin D deficit has been reported to worsen sensitization and allergic manifestations in several different experimental models. However, in clinical situations, contradictory findings have been described concerning the correlation between allergy and vitamin D deficit. The aim of this review was to analyze the close relationships between mast cells and vitamin D, which contribute, through the activation of different molecular or cellular activation pathways, to the determination of bone pathologies and the onset of allergic diseases.

Keywords: mast cell; vitamin D; allergy; osteoporosis; mastocytosis; bone diseases; immune response

1. Introduction

Allergies and osteoporosis are disorders with a great incidence in the overall population and constitute a real danger for community health [1]. Furthermore, several experimental findings have shown a close correlation between the two conditions. Many allergic conditions such as asthma, eczema, chronic respiratory pathologies, and pollen allergy are correlated with fracture risk [2,3], while increased percentages of bone diseases are described in patients with allergic syndromes, in both adults and children [4].

For instance, the connection of asthma with osteoporosis is well established, although up until the recent past, most studies attributed the correlation to the unfavorable effect exerted by protracted steroid therapies on the bone. New findings have instead highlighted the responsibility of the systemic inflammation present in allergic diseases in the osteoporosis onset [5,6], and a correlation between allergies and osteoporosis has been also demonstrated in patients suffering from chronic rhinosinusitis [5].

As for the mechanisms capable of connecting bone diseases and allergic pathologies, a correlation between bone and immune effectors is now recognized, and several relevant communications are seen in the bone milieu, such as the enrolment and growth of T cells [7]. It is also known that bone cells and immune effectors originate from the same precursors in the bone marrow (BM) and that the same compounds can control bone cell's proliferation and activity, immune response, and haematopoiesis [8]. Moreover, the destiny of hematopoietic stem cells is regulated by autocrine and paracrine systems, these phenomena being controlled by bone tissue, osteoclasts, osteoblasts, and immune effectors interacting with each other. Several blood cells, including basophils, mast cells, eosinophils, lymphocytes, and neutrophils contribute to this composite crosstalk, which causes the allergic inflammation. This condition can generate cytokines, reactive oxygen species, chemokines, and lipids leading to a plethora of effects which are able to elicit bone modifications [9].

For instance, the Cysteinyl leukotrienes (CysLTs) are a group of lipid mediators originating from arachidonic acid [10]. The CysLTs receptor (CysLTR1) has a relevant effect in the onset of asthma, and montelukast is a CysLTR1 antagonist employed for the therapy of asthma. An experimentation demonstrated that montelukast effectively inhibits RANKL-caused osteoclast generation and bone damage in vivo [11].

However, one of the main actors of this communication is represented by mast cells (MCs), which play an extremely important role both in the onset of allergic pathologies and in diseases of bone metabolism, and the effect exerted by vitamin D on these cells could constitute a unifying element for apparently heterogeneous alterations (Figure 1).

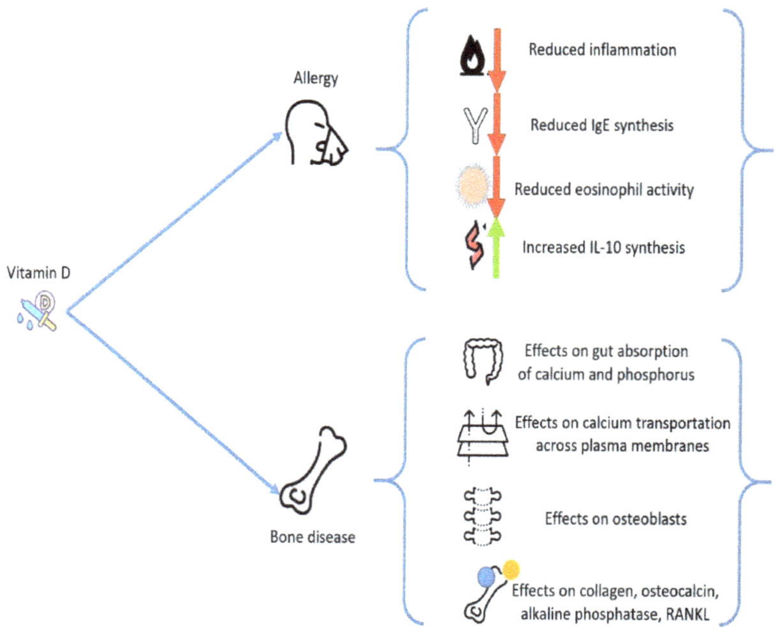

Figure 1. Possible effects of vitamin D on the onset of allergic diseases and on bone metabolism.

The aim of this review was to detect the close connections between mast cells and vitamin D, which contribute, through the activation of different molecular or cellular activation pathways, to the determination of bone pathologies and the onset of allergic diseases.

2. General Considerations on Vitamin D

Vitamin D3 (cholecalciferol) is mainly produced in the epidermis when pre-vitamin D3, the compound originating from 7-dehydrocholesterol after ultraviolet-B (UVB) irradiation of the skin, undertakes thermal isomerization. The transformation of vitamin D3 to its active form 1α,25(OH)2D3, is due to a sequence of hydroxylation procedures, initially caused by liver cytochrome P450 proteins, which can produce the transitional metabolite, 25OHD3 and, subsequently, 25-hydroxyvitamin 1αhydroxylase in the renal tubule to generate 1α,25(OH)2D3. This compound operates by joining to the vitamin D receptor (VDR), which provokes the enrolment of the retinoid X receptor, to generate a heterodimeric structure that affects vitamin D (VD) response factors in the promoter areas of genes. According to the concurrent joining of nuclear co-stimulators or coinhibitory factors, the complex can operate as a ligand-dependent stimulator or an inhibitor of gene transcription [12–14].

VD deficit has been recognized as a main health problem, which is constantly more prevalent [15]. Even though there is no agreement on the ideal concentrations of 25(OH)D, a VD deficit is identified when concentrations of VD less than 50 nmol/L occurs [16].

All organs and cells feature a VDR, including immune and bone cells, skin, brain, gonads, and heart cells. A reduction on VD levels may change the activity of these tissues. This factor explains the extensive effects of VD and justifies why a decrease in VD levels has been related with several different chronic pathologies. Regarding the subject of our analysis, VD has a main action in calcium/phosphate equilibrium and provokes profound consequences on the bone metabolism, beyond having a pivotal effect as an anti-inflammatory mediator. VD deficit increases the possible occurrence of osteoporosis and several other pathologies that present alterations of bone metabolisms, such as hematological malignancies, bone marrow transplantation, inflammatory bowel diseases, and endocrinological diseases [17].

Furthermore, VD modifies relevant activities of the immune system and may change the development of immune-mediated diseases, including allergies and autoimmune diseases. VD operates by steering T lymphocytes to Th2 polarization and blocking Th1 and Th17 function and growth. The stimulation of T regulatory cells (Treg) might be its principal immunological function [18]. Numerous reports indicate a VD favorable action on diseases correlated to hyperstimulation of Th1 cells, such as psoriasis, multiple sclerosis, rheumatoid arthritis, and type 1 diabetes [19]. In allergic pathologies, where Th2 cells have a central effect, VD has a more complex effect. However, several experimentations displayed a positive action on the progression of allergic conditions, although the causal mechanisms have not always been fully explained [20].

3. Vitamin D and Allergies

Epidemiological data have also indicated the existence of a correlation of VD reduction with several conditions including allergies. The relationship of VD deficit and asthma is described, and several studies stated a connection with disease course worsening and a worse outcome. It is possible that VD, by improving the inflammation state, can decrease the rate of respiratory sepsis and exacerbations [21,22]. Although the correlation between VD concentrations and allergies has been questioned in a cross-sectional analysis including only allergic subjects [23], an effect of VD in the eosinophil activities is demonstrated. VD decreases the immunoglobulin E (IgE) synthesis and increases expression of interleukin-10 [24].

Other studies confirmed the correlation of VD with allergic diseases [25,26]. Some experimentations were conducted to evaluate the effect of VD in chronic urticaria (CU), allergic contact dermatitis (ACD), and atopic dermatitis (AD) [27,28], and a relationship between VD deficit with ACD and AD severity was reported [29,30], while a work proved a relevant difference between CSU subjects and controls in blood concentrations of VD [31].

An interesting analysis has reported that the period of birth and UBV exposure is correlated to the incidence of food allergy (FA). The concentrations of VD generated from skin after UBV might justify this correlation, while maternal or VD deficiency occurring in the first years of life also remarkably predisposed the FA onset in an experimental animal model [32,33].

Still in the framework of FA, employing results obtained from a nutritional investigation, it was stated that allergic sensitization to several different allergens, including foods, was more frequent in young subjects with a 25(OH)D deficit, while no relevant correlation was found between VD concentrations and FA in adults [34]. Although, a different study performed by Beak et al. reported that reduced concentrations of VD might be correlated to polysensitization of nutrition allergens [35]. Therefore, the correlation between VD and FA is still debatable.

However, other findings support the existence of a tight relationship between VD and allergies, and the beneficial action of VD supplementation in CSU subjects has been proved in several small clinical experimentations [36,37]. In a prospective research, patients suffering from CSU were treated with low or high vitamin D3 supplement for 12 weeks. All patients demonstrated a significantly decreased urticaria severity score, but the reduction was greater in patients treated with a higher VD dosage at week six. The medication score was also remarkably decreased, without distinguishing the two groups [38].

We report in the table some of the works that confirm the existence of a relationship between VD levels and allergic diseases [27,34,37–46], (Table 1).

Table 1. Effects of Vitamin D on allergies.

Allergic Disease	Correlation with Vitamin D	Ref.
Atopic disease	Higher values of 25OH-D3 in patients with mild disease compared with patients with moderate or severe disease	[27]
	Reduced VD levels in Atopic Disease patients	[39]
	Vitamin D supplementation during winter months had favorable effects on Atopic Disease symptoms	[40]
Food Allergy	Presence of 25OH-D3 serum values of less than 15 ng/mL	[34]
Chronic Urticaria	Reduced Urticaria Symptoms Severity scores after D3 supplementation	[37]
Chronic Spontaneous urticaria	Reduced total urticaria score after vitamin D3 administration	[38]
Asthma	Presence of 25OH-D3 serum values of less than 20 ng/mL (3.4%), and of 20–30 ng/mL (24.6%)	[41]
	Inverse correlation between maternal 25OH-D3 values and inhaled steroids	[42]
	Vitamin D supplementation during winter months reduced the frequency of asthma attacks	[43,44]
Childhood wheezing	Correlation between high maternal 25OH-D3 values with reduced childhood wheezing	[45,46]

4. Vitamin D and Immune Response

Several other relevant connections have been reported between blood cells, both lymphoid and myeloid cells, and VD concentrations. For instance, B cells can produce VD [47], while naïve T cells grown with VD-primed B cells demonstrated decreased proliferation, provoked by the presence of CD86 on B cells [48].

As reported above, VD has a main effect in IgE provoked responses, and it was displayed that IgE serum concentrations are enhanced in VDR-knockout animals [49]. Similarly, VD reduced IgE generation by B lymphocytes, and the IgE reaction in a type 1 allergy animal model can be altered by employing a VDR agonist [50].

Furthermore, VD can act on other effectors of the immune system, and Szeles et al. have stated that an increase of 25(OH)D provokes dendritic cells (DCs) to switch on VD related genes, and stimulation of VDRs by VD resets the DCs to turn out to be tolerogenic [51]. Remarkably, when VD was added to monocyte, DCs were less mature, presenting different amounts of MHC class II molecules and transforming CD4+ T cells to IL-10-producing Tregs [52]. Thus, VD not only produces a repressive consequence on DC development but also instructs the DCs to stimulate Tregs to generate IL-10 [53]. Almerighi et al. have confirmed that VD reduces inflammation caused by CD40L and increases IL-10 generation by CD4+ T cells [54], stimulating the expansion of Tregs presenting forkhead box P3 and cytotoxic T-lymphocyte-associated protein 4 [55] (Figure 2).

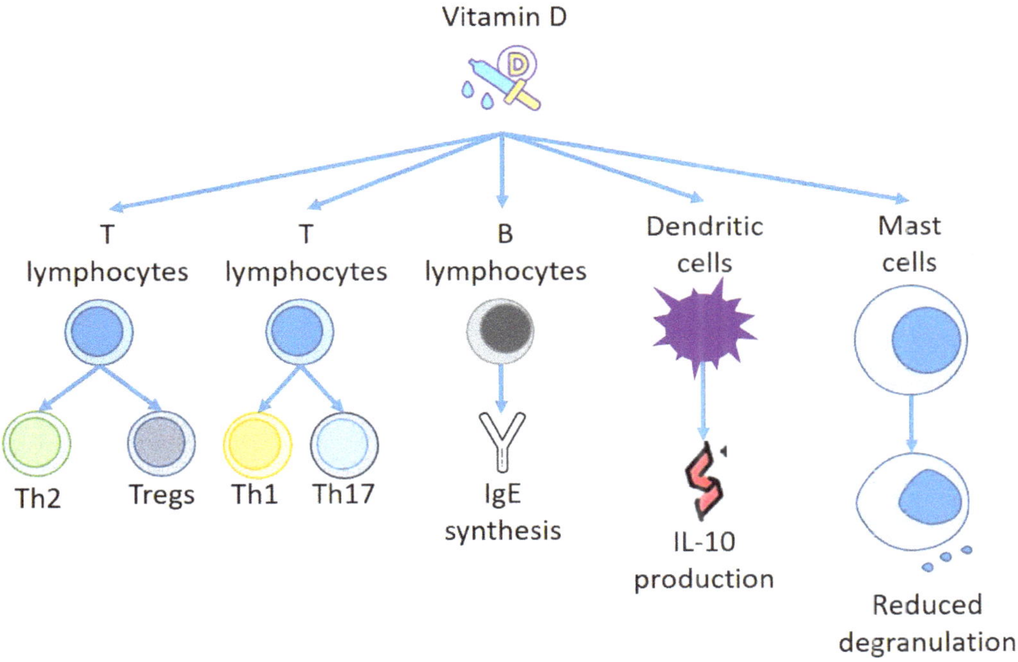

Figure 2. Effects of Vitamin D on immune effectors.

5. Vitamin D and Mast Cells: Effects on Allergies

VD also seems to have a role in allergic diseases and bone pathologies mediated by Th2 cells, via the IL-31/IL-33 axis [56], which is another new area to investigate the intricate overlapping mechanisms that connect allergies and osteoporosis. IL-31 is a cytokine that is able to provoke inflammation, which was proposed as a marker of tissue remodeling and different allergic and immunologic pathologies [56]. It is released by Th2 cells and, in smaller amounts, also by DCs and MCs. IL-31 stimulates eosinophils and fibroblasts, and its receptor is present in skin and endothelium. It is correlated with the onset of itch and chronic skin inflammation. An increased amount of IL-31 was reported in the skin and serum of patients suffering from AD, CSU, contact dermatitis, prurigo nodularis, cutaneous lymphoma, and mastocytosis, correlating with disease severity [57,58]. It is therefore interesting that increased concentrations of serum IL-31 are present in postmenopausal women with a reduction of bone mineral density (BMD), although there is no correlation with the degree of osteoporosis [59].

Mast cells are myeloid cells that transfer into practically all tissues, where they perform tissue–specialized evolution. They are a constituent of connective tissue and are numerous in areas such as the gut, mucous membranes, and skin [60,61], near to vessels and nerves [62]. This tactical settlement and the mediators generated from MCs explain their ability to quickly modify their milieu, and to control other forms of inflammatory cells [63,64].

MCs have a relevant effect in immediate hypersensitivity reactions, but also in late phase responses and innate immune response, by generating a large number of different mediators either from storage places in their granules or by delivering substances which are able to regulate different signaling pathways after adequate stimulation [65].

In allergic reactions, IgE joins to the IgE receptor to produce complexes on the cellular membrane of MCs to induce MC sensitization. After new contact with particular antigens, they combine with the IgE/FceRI complex to stimulate MCs [66]. Moreover, MCs also can be stimulated by temperature modifications and microbial factors [67,68].

However, beyond allergic diseases, MCs are involved in the genesis of a massive variety of pathological conditions including chronic inflammation, autoimmunity, and cancer [69].

As for the relationship between MCs and VD, the cells discharged several mediators in a VD-lacking milieu without the need of any activators. Presence of calcitriol in the culture medium increased the number of VDRs in the MCs, and VDRs provoked complexes with Lyn in MCs to block the connection of Lyn to MyD88 and to the β chain of FcεRI, which diminished the amounts of NF-kB and MAPK and reduced the phosphorylation of Syk. Moreover, VDRs connected to the promoter of TNF-α reduce the acetylation of RNA polymerase II and histone H3/H4, reducing the production of TNF-α in MCs. These findings make evident that VD is necessary to preserve the steadiness of MCs, whereas the deficit of VD provokes the stimulation of MCs [70,71].

This has been demonstrated under numerous experimental conditions. It is notorious that the release of granules and the discharge of histamine from MCs are involved in the genesis of urticaria [72,73]. VD has been suggested for this therapy and for the one concerning other allergic pathology [74,75], as MCs own the VDR capable of blocking degranulation of compounds provoked by IgE [76].

Finally, it may be important to consider how the strong correlations between MCs and VD can aid to clarify some contradictory actions provoked by MCs. Although MCs were once believed to operate essentially as cells which were able to induce inflammation that can provoke allergic responses and inflammation induced by exogenic factors such as UV irradiation [63,77], novel findings suggest that, in specific conditions, MCs can reduce cellular damage and decrease inflammation provoked by UVB irradiation [78]. Even though most UVB rays do not enter the dermis, the UVB–supported stimulation of dermal MCs is believed to be obtained through a nerve growth factor, that originated from the epidermis [79], or by nervous sensory C fibers that have been stimulated by cis-urocanic acid [80]. This secondary mechanism of MC stimulation seems to participate in the UVB-provoked systemic immunosuppression in animals exposed to an acute single dose of UVB irradiation [79]. The intimate mechanism of this condition could be constituted by the production of specific cytokines, including IL-10 (Figure 3).

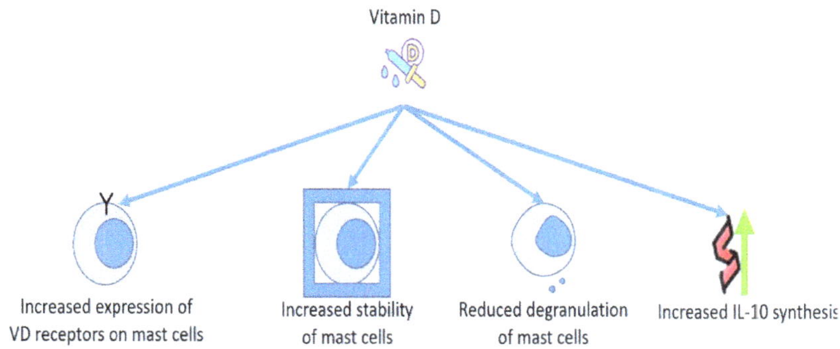

Figure 3. Effects of Vitamin D on mast cells.

MC's generation of IL-10 can reduce the skin damage provoked by chronic UVB irradiation. Even if the specific mechanism that induces MC IL-10 generation in this condition is undetermined, it is well-known that exposure of the skin to UVB fosters the generation of 1 alpha,25[OH]2D3. A study displayed that VD could increase IL-10 mRNA generation and stimulate IL-10 delivery in animal MCs in vitro. A different experimentation evaluated the effect of VD and MCs VDR expression in UVB irradiated skin in vivo. Researchers engrafted the skin of genetically mast cell-deficient WBB6F(1)-Kit(W/W-v) mice with BM-originating MCs isolated from C57BL/6 wild-type or VDR-/- animals. MC-supported reduction of the inflammation correlated with chronic UVB irradiation of the skin in the KitW/W-v animal's expression of VDRs by the adoptively transferred MCs. As VD can produce IL-10, these results indicate that VD/VDR-dependent stimulation of IL-10 generation by skin MCs can participate to the MCs capability to reduce skin inflammation after chronic UVB irradiation [71].

These findings suggest that stimulating the anti-inflammatory effects of MCs by adding VD might be a new strategy for decreasing tissue injury and inflammation in several different pathological conditions.

However, the relationship between mast cells and VD could be even more relevant. Yip et al. demonstrated that MCs could transform 25OHD3 to 1α, 25(OH)2D3 through 25-hydroxyvitamin D-1α–hydroxylase, and that these VD3 derivatives reduced the MCs production of pro-inflammatory mediators in a VDR-dependent modality [81]. Furthermore, when VD3 metabolites were epicutaneously employed, a remarkable decrease in skin inflammation occurred. This effect requires the presence of MCs-VDRs and MCs-CYP27B1. These results provide a mechanistic justification for the anti-inflammatory action of VD on MCs' activity by showing that MCs can metabolize 25OHD3 to inhibit IgE-mediated MCs stimulation in both in vivo and in vitro models.

6. Mast Cells, Vitamin D, and Bone Metabolism

The relationship between MCs and VD goes far beyond the possibility of VD to influence the activity of MCs in allergic manifestations and both have been shown to have a role in determining bone pathology. In fact, MCs generate mediators that alter bone metabolism.

It has been known for a long time that MCs may have a role in fracture repairing [82]. Histology analysis of fractures in animal models demonstrated that, in the first fifteen days, MCs are present in the surrounding area of blood vessels and in the vascularized areas growing into subperiosteal callus [83]. This finding suggests that MCs might be implicated in the angiogenic process and in the metabolism of extracellular matrix in the initial phases of fracture repairing.

However, it was reported that MCs might reduce bone alteration repairing by activating arteriogenesis, and that the successful effect of recombinant parathyroid hormone treatment on bone healing is due to the reduction of arteriogenesis secondary to MCs inhibition [84].

Some of the effects exerted by MCs on bone metabolism may be due to the action of VD. It has been reported that MCs granules include VD, which is not produced inside the granules. VD produced by the keratinocyte transits into the intercellular milieu, and travels toward the basement membrane; at this level VD is absorbed by MCs, where it is deposited within its granules [85,86].

However, the mediators contained or produced by MCs are several hundred, and the overall effect of these cells on bone metabolism can be deeply different depending on the mediator considered and the different experimental conditions.

7. Mastocytosis, Allergy and Bone Alterations

The role played by MCs in allergology and on bone metabolism and any correlations with VD are highlighted in what constitutes the most serious alteration of MCs, mastocytosis, a hemopoietic malignancy characterized by the clonal proliferation of these cells and their accrual in different organs and tissues. In more than 80% of subjects with systemic mastocytosis (SM), a somatic point genetic mutation in KIT at codon 816 is discovered [87,88].

Diagnosis is performed by employing the major criterion of multifocal clusters of altered MCs in BM and the minor criteria of the occurrence of the KIT D816V mutation, increased levels of serum tryptase, and altered MC CD25 expression [89]. Cutaneous mastocytosis is most common in infancy, while systemic mastocytosis concerns adult patients and extracutaneous tissues, with or without skin participation.

The WHO classified SM into: indolent SM (ISM); smouldering SM (SSM); SM with an associated hematologic neoplasm (SM-AHN); aggressive SM (ASM); and MC leukemia (MCL). Moreover, MC sarcoma (MCS) is an extremely severe form of mastocytosis identified by a sarcoma-like proliferation of abnormal, clonal MCs, without a systemic diffusion. The outcome in these subjects is poor, and patients evolve to MCL within a brief period [90,91].

ISM is the most common form of the disease, and its symptomatology is essentially due to the release of MC mediators, discharge, and the onset of allergic manifestations. As far as the bone alterations, ISM has been identified as a risk element for reduced BMD. Considering the conventional WHO criteria [92], osteoporosis in ISM subjects goes from 18 to 28% [93,94]. According to the more recent parameters suggested by the International Society for Clinical Bone Densitometry [95], mastocytosis-correlated reduction of BMD was recognized in about 9% of women and in 28% of men with the disease [96].

However, in all subjects suffering from mastocytosis, irrespective of sex, age, or disease form, an osteopathy can occur, and VD deficit [97], osteopenia, and osteoporosis are commonly found in SM subjects [98–100].

In a group of patients suffering from ISM without osteoporosis, Artuso et al. studied BMD change at the lumbar spine and proximal hip before and after VD supplementation [101]. Furthermore, they investigated the possible relationships between BMD changes and serum tryptase concentrations. The presence of a VD deficit was reported in more than 70% of ISM subjects. After 30 months with VD (or calcium supplement when necessary), they reported a 2.1% increase in BMD at the lumbar spine, while no variations were detected at hip. However, after 30 months, about 60% of subjects presented VD concentrations that were still lower than recommended [101].

Other experimental data collected on animal models seem to confirm the connection between mast cell and VD during mastocytosis. Cutaneous mast cell tumors (MCT) arise more commonly in dogs than in other animals, and reduced concentrations of VD might be a risk element for MCT [102,103]. Calcitriol is recognized to possess anti-proliferative actions on several canine neoplastic cells, including the C2 cell line originated from a cutaneous MCT [104]. Remarkably, VDR is extensively present in both inflammatory and clonal canine MCs, but the recurrence rate of VDR expression is higher in tumoral MCs [105].

It is likely that the bone changes present in patients with mastocytosis have a multifactorial origin.

First, ISM-correlated osteoporosis is sustained by an increased osteoclastic activity. This supposition is also supported by the fact that anti-resorption substances are efficacious in drugs and effective in managing such conditions [94]. A recent study displayed a relationship between tryptase, osteoprotegerin, and C-terminal telopeptide of type 1 collagen, in subjects suffering from severe SM variants [26,106]. Furthermore, RANK–RANKL signaling leads the osteoclastic activity and MCs express RANK ligand, activating the osteoclasts, causing bone resorption, and osteoporosis [107]. The increase in osteoclast activity by the RANK ligand enhancement is also able to provoke the osteolytic alterations [108]. This mechanism is due to the compounds delivered by MCs. Different cytokines, such as IL-1, IL-6, and TNF-alpha, as well as tryptase, histamine, heparin, all are discharged by MCs, with direct consequences on osteoclasts and osteoblasts [100]. Indeed, these substances stimulate osteoclast function provoking bone loss and inducing osteoporosis and osteopenia, while histamine has been reported to increase bone growth causing osteosclerosis [109] (Figure 4).

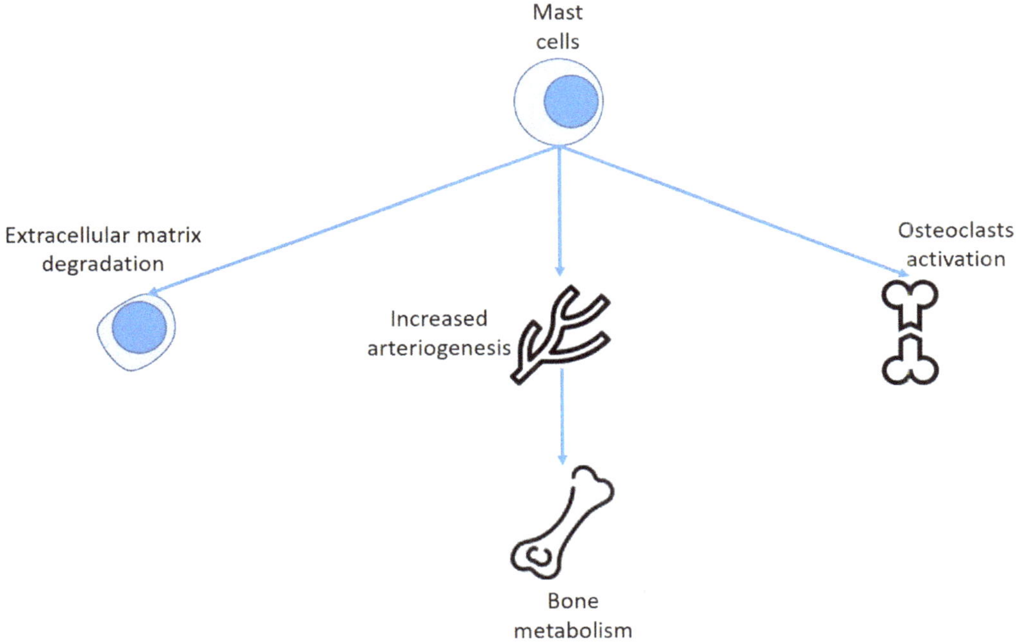

Figure 4. Effect of mast cells on bone metabolism.

Moreover, histamine working via many different receptors is answerable for the allergic effects such as pruritus, and the consequences on bronchial and gastrointestinal smooth muscle cells. Therefore, bone changes are the overall product of the chymotryptic proteases, cytokines, RANKL, fibroblast growth factor family proteins, hormonal factors, and histamine [96,110–112].

Another mechanism that appears to correlate mast cells, VD, allergy, and bone disease could be represented by their effects on the vascular endothelial growth factor (VEGF). VEGF is a substance able to modify vascular permeability increasing the generation of nitric oxide. Several findings demonstrated that VEGF has a fundamental action in allergic pathologies including CSU and is a blood marker for CSU identification [113]. MCs are one of the main producers of VEGF. Although VEGF is essential in bone repair and remodeling, VEGF operates as a powerful chemoattractant for osteoclasts. Moreover, VEGF production is enhanced during osteoclast differentiation via NF-κB stimulation of HIF-1α. Furthermore, osteoclasts express VEGF receptors and VEGF directly increases osteoclast survival and osteoclastic bone resorption. An experimentation stated that VD binding protein (VDBP) was increased in CSU subjects with respect to normal controls. Furthermore, sera of CSU subjects stimulated the generation of VEGF in MCs via PI3K/Akt/p38 MAPK/HIF-1α axis in an IgE-supported modality, and VD reduced the production of VEGF by blocking this axis [114]. Together, these findings indicate a possible beneficial effect exerted by the administration of VD in reducing the onset of osteoporosis in allergic patients.

In conclusion, when MCs discharge their mediators, several metabolic alterations may occur, among which are bone remodeling and osteolysis. This is a certain fact in the case of myeloproliferative neoplasms such as mastocytosis but must be further evaluated in allergic conditions. Bone remodeling is modified by MCS mediators that alter the equilibrium between bone destroying and bone generation. Mastocytosis is accompanying with great rate of recurrence to an abnormal bone mineralization [115], but bone modifications are also described in asthma and other allergic diseases. However, there is not enough data about a conclusive responsibility of MCs in osteoporosis in allergic pathologies, yet. MCs are too numerous in mastocytosis and too functioning in allergic diseases, and this indicates that the effect on bone metabolism might be analogous.

Bone alterations correlated to mastocytosis, and allergies should be managed with adequate VD and calcium supplement. Other studies are needed to evaluate the possibility of treating the bone changes in these patients, using bisphosphonates or denosumab [116].

8. Mast Cells, Vitamin D and Skin Diseases

The existence of MCs is a characteristic feature of different skin pathologies and tumors, and MCs can be involved in the onset and progression of tumors via the effects exerted on mitogenesis and angiogenesis, the effects on the extracellular matrix, and the effects on the immune response [117,118]. It has been theorized that MCs may have a different phenotype in several pathologic conditions and may present an immunosuppressive set-up in skin tumors, but a proinflammatory set-up in chronic skin inflammation [110]. However, the reason and the mechanisms able to cause a different phenotype are not well clarified. Exposure to UV can induce an immunosuppressive set-up in the skin, and VD might be involved in this switch [119]. In fact, as reported above, an interesting novel mechanism has been reported in animal models, in which UVB caused the generation of vitamin D3, provoking the expression of VDR in MCs and synthesis of the immunosuppressive IL-10 [120]. The leukemic HMC-1 MC line has been reported to present VDR mRNA, and isolated skin MCs have also been described to present VDR mRNA, proposing that this process could also be important in the skin [121]. However, a different mechanism may imply the MCs synthesis of vitamin D3 derivative end products, as MCs present CYP27A1 and CYP27B1 in several skin diseases [122], and these hydrolases may stimulate the synthesis of calcidiol from vitamin D3 and calcitriol from calcidiol [123]. These vitamin D3 derivatives can work as a VDR agonist ligand [124]. CYP27A1 was reported to be notably expressed in MCs BCC and SCC [122] (Figure 5).

However, the situation is probably much more complex. In fact, VDR stimulation provokes the production of CYP24A1, a hydroxylase that can block vitamin D3 derivatives. To evaluate immunoreactivity to VDR and CYP24A1 in MCs, skin biopsies were collected from the non-lesional and lesional skin of patients with BCC, SCC, actinic keratosis (AK), and psoriasis. In non-lesional skin biopsies, only 0.5–2.9% of the MCs presented CYP24A1, however, the rate of MCs presenting CYP24A1 was remarkably increased in lesional skin of SCC, BCC, and AK. In contrast to human skin, LAD2 MCs isolated from a subject with mast cell sarcoma/leukaemia demonstrated that around 34% and 6.5% of the cells were, in turn, positive for VDR and CYP24A1. Thus, while a very slight percentage of MCs in human skin present VDR and CYP24A1, the amount of MCs presenting CYP24A1 in keratinocyte skin tumors is increased [125]. The stimulation of VDR by calcitriol induces an increase of CYP24A1, in turn supposed to cause a negative regulation to provoke a reduction of calcitriol effect.

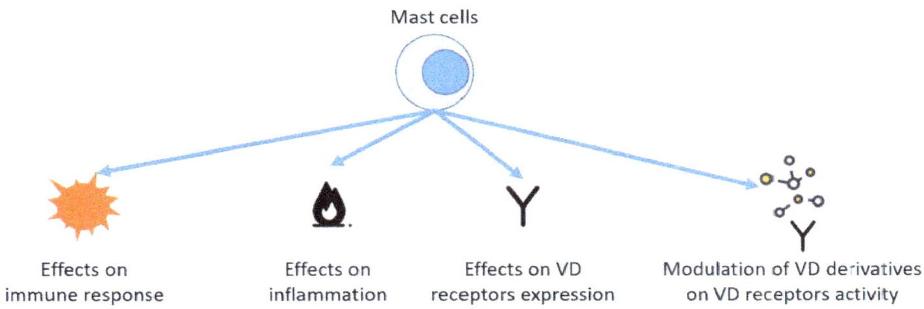

Figure 5. Effects of mast cells on skin disease.

9. Conclusions

Experimental and epidemiological results indicate that VD deficits are correlated with the genesis of allergic diseases. In fact, VD changes immune response, MCS proliferation, and function. It also seems certain, from the data presented so far, that VD and MCs are involved in the control of bone metabolism and can intervene in the determinism of bone pathologies.

However, we are far from understanding the real mechanisms that regulate the complex interactions between MCs and VD, and sometimes contradictory data are present in the literature.

For instance, not only a VD deficiency, but also increased VD concentrations have been correlated to an increase in IgE production [126,127]. Cairncross et al. stated that increased 25(OH)D levels were correlated with a greater risk of FA in children [128], and a different study demonstrated that CD1 WT mice treated with 1α,25(OH)2D3 presented an inflammation of the skin similar to atopic dermatitis. A high dosage of 1α,25(OH)2D3 remarkably increased thymic stromal lymphopoietin mRNA concentrations. However, the effects were different in a dissimilar experimental animal model. This difference might be attributed to the heterogeneous genetic conditions of the animals [129]. Thus, both deficient and excessive amounts of vitamin D are potential risk factors associated with allergic disease.

Therefore, it is likely that different climate, seasonal, and genetic elements can affect the evaluation of the correlations occurring between MCs and VD and the consequences of their reciprocal effects on allergic diseases and bone metabolism pathologies. Further experiments will make it possible to definitively clarify these relationships and to use the administration of VD analogues for more adequate treatment of these pathologies and mast cell activation states.

Author Contributions: Conceptualization, G.M., A.A. and S.G; methodology, A.T., C.M., L.R.; software, A.T.; formal analysis, A.A., A.T.; writing—original draft preparation, A.A.; writing—review and editing, G.M., A.A., S.G., A.T., C.M., L.R. All authors have read and agreed to the published version of the manuscript.

Funding: This research received no external funding.

Institutional Review Board Statement: Not applicable.

Informed Consent Statement: Not applicable.

Data Availability Statement: Not applicable.

Conflicts of Interest: The authors declare no conflict of interest.

References

1. De Martinis, M.; Sirufo, M.M.; Viscido, A.; Ginaldi, L. Food Allergy Insights: A Changing Landscape. *Arch. Immunol. Ther. Exp* **2020**, *68*, 8. [CrossRef] [PubMed]
2. Garg, N.; Silverberg, J.I. Association between eczema and increased fracture and bone or joint injury in adults a us population based study. *JAMA Dermatol.* **2015**, *151*, 33–41. [CrossRef]
3. Chen, Y.W.; Ramsook, A.H.; Coxson, H.O.; Bon, J.; Reid, W.D. Prevalence and Risk Factors for Osteoporosis in Individuals with COPD: A Systematic Review and Meta-analysis. *Chest* **2019**, *156*, 1092–1110. [CrossRef]
4. Barrick, B.J.; Jalan, S.; Tollefson, M.M.; Milbrandt, T.A.; Larson, A.N.; Rank, M.A.; Lohse, C.M.; Davis, D.M.R. Associations of self-reported allergic diseases and musculoskeletal problems in children: A US population-based study. *Ann. Allergy Asthm Immunol.* **2017**, *119*, 170–176. [CrossRef]
5. Aljubran, S.A.; Whelan, G.J.; Glaum, M.C.; Lockey, R.F. Osteoporosis in the at-risk asthmatic. *Allergy* **2014**, *69*, 1429–143 [CrossRef]
6. Jung, J.-W.; Kang, H.-R.; Kim, J.-Y.; Lee, S.-H.; Kim, S.S.; Cho, S.H. Are asthmatic patients prone to bone loss? *Ann. Allergy Asthr Immunol.* **2014**, *112*, 426–431. [CrossRef] [PubMed]
7. Choi, H.G.; Kong, I.G. Association between chronic rhinosinusitis and osteoporosis: A case-control study using a national sampl color. *Int. Forum Allergy Rhinol.* **2019**, *9*, 1010–1016. [CrossRef] [PubMed]
8. Zupan, J.; Jeras, M.; Marc, J. Osteoimmunology and the influence of pro-inflammatory cytokines on osteoclasts. *Biochem. Me* **2013**, *23*, 43–63. [CrossRef]
9. Ponzetti, M.; Rucci, N. Updates on Osteoimmunology: What's New on the Cross-Talk between Bone and Immune System. *Fron Endocrinol.* **2019**, *10*, 236. [CrossRef]
10. Naik, S.R.; Wala, S.M. Inflammation, allergy and asthma, complex immune origin diseases: Mechanisms and therapeutic agent *Recent Patents Inflamm. Allergy Drug Discov.* **2013**, *7*, 62–95. [CrossRef]
11. Kang, J.; Lim, H.; Lee, D.; Yim, M. Montelukast inhibits RANKL-induced osteoclast formation and bone loss via CysLTR1 an P2Y12. *Mol. Med. Rep.* **2018**, *18*, 2387–2398. [CrossRef]
12. Bouillon, R.; Carmeliet, G.; Verlinden, L.; van Etten, E.; Verstuyf, A.; Luderer, H.F.; Lieben, L.; Mathieu, C.; Demay, M. Vitamin and human health: Lessons from vitamin D receptor null mice. *Endocr Rev.* **2008**, *29*, 726–776. [CrossRef] [PubMed]
13. Pike, J.W. Genome-wide principles of gene regulation by the vitamin D receptor and its activating ligand. *Mol. Cell. Endocrinc* **2011**, *347*, 3–10. [CrossRef] [PubMed]
14. Haussler, M.R.; Jurutka, P.W.; Mizwicki, M.; Norman, A.W. Vitamin D receptor (VDR)-mediated actions of 1alpha,25(OH)(2)vitam D(3): Genomic and non-genomic mechanisms. *Best Pract. Res. Clin. Endocrinol. Metab.* **2011**, *25*, 543–559. [CrossRef] [PubMed]
15. Elsori, D.H.; Hammoud, M.S. Vitamin D deficiency in mothers, neonates and children. *J. Steroid Biochem. Mol. Biol.* **2018**, *17* 195–199. [CrossRef]
16. Holick, M. Vitamin D deficiency. *N. Engl. J. Med.* **2007**, *357*, 266–281. [CrossRef] [PubMed]
17. De Martinis, M.; Allegra, A.; Sirufo, M.M.; Tonacci, A.; Pioggia, G.; Raggiunti, M.; Ginaldi, L.; Gangemi, S. Vitamin D Deficienc Osteoporosis and Effect on Autoimmune Diseases and Hematopoiesis: A Review. *Int. J. Mol. Sci.* **2021**, *22*, 8855. [CrossRef]
18. Bivona, G.; Agnello, L.; Ciaccio, M. The immunological implication of the new vitamin D metabolism. *Cent. Eur. J. Immunol.* **201** *43*, 331–334. [CrossRef] [PubMed]
19. Ciccarelli, F.; De Martinis, M.; Sirufo, M.M.; Ginaldi, L. Psoriasis Induced by Anti-Tumor Necrosis Factor Alpha Agents: Comprehensive Review of the Literature. *Acta Dermatovenerol. Croat. ADC* **2016**, *24*, 169–174. [PubMed]
20. Muehleisen, B.; Gallo, R.L. Vitamin D in allergic disease: Shedding light on a complex problem. *J. Allergy Clin. Immunol.* **201** *131*, 324–329. [CrossRef] [PubMed]
21. Bouillon, R.; Marcocci, C.; Carmeliet, G.; Bikle, D.; White, J.H.; Dawson-Hughes, B.; Lips, P.; Munns, C.F.; Lazaretti-Castro, M Giustina, A.; et al. Skeletal and Extraskeletal Actions of Vitamin D: Current Evidence and Outstanding Questions. *Endocr. Re* **2019**, *40*, 1109–1151. [CrossRef] [PubMed]
22. Marino, R.; Misra, M. Extra-Skeletal Effects of Vitamin D. *Nutrients* **2019**, *11*, 1460. [CrossRef] [PubMed]
23. Lombardi, C.; Passalacqua, G. Italian Vitamin D Allergy Group Vitamin D levels and allergic diseases. An Italian cross-section multicenter survey. *Eur. Ann. Allergy Clin. Immunol.* **2017**, *49*, 75–79.
24. Souto Filho, J.T.D.; de Andrade, A.S.; Ribeiro, F.M.; Alves, P.A.S.; Simonini, V.R.F. Impact of vitamin D deficiency on increase blood eosinophil counts. *Hematol. Oncol. Stem Cell Ther.* **2018**, *11*, 25–29. [CrossRef] [PubMed]
25. Alyasin, S.; Momen, T.; Kashef, S.; Alipour, A.; Amin, R. The relationship between serum 25-hydroxyvitamin D levels and asthm in children. *Allergy Asthma Immunol. Res.* **2011**, *3*, 251–255. [CrossRef]
26. Arshi, S.; Fallahpour, M.; Nabavi, M.; Bemanian, M.H.; Javad-Mousavi, S.A.; Nojomi, M.; Esmaeilzadeh, H.; Molatefi, R.; Rekab M.; Jalali, F.; et al. The effects of vitamin D supplementation on airway functions in mild to moderate persistent asthmal. *An Allergy Asthma Immunol.* **2014**, *113*, 404–409. [CrossRef]
27. Peroni, D.; Piacentini, G.; Cametti, E.; Chinellato, I.; Boner, A. Correlation between serum 25-hydroxyvitamin D levels ar severity of atopic dermatitis in children. *Br. J. Dermatol.* **2011**, *164*, 1078–1082. [CrossRef] [PubMed]
28. Dogru, M. Is vitamin D level associated with the natural course of atopic dermatitis? *Allergol. Immunopathol.* **2018**, *46*, 546–55 [CrossRef]

29. Grzanka, A.; Machura, E.; Mazur, B.; Misiolek, M.; Jochem, J.; Kasperski, J.; Kasperska-Zajac, A. Relationship between vitamin D status and the inflammatory state in patients with chronic spontaneous urticarial. *J. Inflamm.* **2014**, *11*, 24484740. [CrossRef] [PubMed]
30. Quirk, S.K.; Rainwater, E.; Shure, A.K.; Agrawal, D.K. Vitamin D in atopic dermatitis, chronic urticaria and allergic contact dermatitis. *Exp. Rev. Clin. Immunol.* **2016**, *12*, 839–847. [CrossRef] [PubMed]
31. Kolkhir, P.; André, F.; Church, M.K.; Maurer, M.; Metz, M. Potential blood biomarkers in chronic spontaneous urticaria. *Clin. Exp. Allergy* **2017**, *47*, 19–36. [CrossRef] [PubMed]
32. Heine, G.; Tabeling, C.; Hartmann, B.; Gonzalez Calera, C.R.; Kuhl, A.A.; Lindner, J.; Radbruch, A.; Witzenrath, M.; Worm, M. 25-hydroxvitamin D3 promotes the long-term effect of specific immunotherapy in a murine allergy model. *J. Immunol.* **2014**, *193*, 1017–1023. [CrossRef] [PubMed]
33. Wu, J.; Zhong, Y.; Shen, X.; Yang, K.; Cai, W. Maternal and early-life vitamin D deficiency enhances allergic reaction in an ovalbumin-sensitized BALB/c mouse model. *Food Nutr. Res.* **2018**, *62*, 1401. [CrossRef] [PubMed]
34. Sharief, S.; Jariwala, S.; Kumar, J.; Muntner, P.; Melamed, M.L. Vitamin D levels and food and environmental allergies in the United States: Results from the national health and nutrition examination survey 2005–2006. *J Allergy Clin. Immunol.* **2011**, *127*, 1195–1202. [CrossRef] [PubMed]
35. Baek, J.H.; Shin, Y.H.; Chung, I.H.; Kim, H.J.; Yoo, E.G.; Yoon, J.W.; Jee, H.M.; Chang, Y.E.; Han, M.Y. The link between serum vitamin D level, sensitization to food allergens, and the severity of atopic dermatitis in infancy. *J. Pediatr.* **2014**, *165*, 849–854.e1. [CrossRef] [PubMed]
36. Yepes-Nuñez, J.J.; Brożek, J.L.; Fiocchi, A.; Pawankar, R.; Cuello-García, C.; Zhang, Y.; Morgano, G.P.; Agarwal, A.; Gandhi, S.; Terracciano, L.; et al. Vitamin D supplementation in primary allergy prevention: Systematic review of randomized and non-randomized studies. *Allergy* **2018**, *73*, 37–49. [CrossRef] [PubMed]
37. Rorie, A.; Goldner, W.S.; Lyden, E.; Poole, J.A. Beneficial role for supplemental vitamin D3 treatment in chronic urticaria: A randomized study. *Ann. Allergy Asthma Immunol.* **2014**, *112*, 376–382. [CrossRef] [PubMed]
38. Nabavizadeh, S.H.; Alyasin, S.; Esmaeilzadeh, H.; Mosavat, F.; Ebrahimi, N. The effect of vitamin D add-on therapy on the improvement of quality of life and clinical symptoms of patients with chronic spontaneous urticaria. *Asian Pac. J. Allergy Immunol.* **2020**. [CrossRef]
39. Vahavihu, K.; Ala-Houhala, M.; Peric, M.; Karisola, P.; Kautiainen, H.; Hasan, T.; Snellman, E.; Alenius, H.; Schauber, J.; Reunala, T. Narrowband ultraviolet B treatment improves vitamin D balance and alters antimicrobial peptide expression in skin lesions of psoriasis and atopic dermatitis. *Br. J. Dermatol.* **2010**, *163*, 321–328. [CrossRef]
40. Sidbury, R.; Sullivan, A.F.; Thadhani, R.I.; Camargo, C.A., Jr. Randomized controlled trial of vitamin D supplementation for winter-related atopic dermatitis in Boston: A pilot study. *Br. J. Dermatol.* **2008**, *159*, 245–247. [CrossRef]
41. Brehm, J.M.; Celedon, J.C.; Soto-Quiros, M.E.; Avila, L.; Hunninghake, G.M.; Forno, E.; Laskey, D.; Sylvia, J.S.; Hollis, B.W.; Weiss, S.T.; et al. Serum vitamin D levels and markers of severity of childhood asthma in Costa Rica. *Am. J. Respir. Crit. Care Med.* **2009**, *179*, 765–771. [CrossRef] [PubMed]
42. Goleva, E.; Searing, D.A.; Jackson, L.P.; Richers, B.N.; Leung, D.Y. Steroid requirements and immune associations with vitamin D are stronger in children than adults with asthma. *J. Allergy Clin. Immunol.* **2012**, *129*, 1243–1251. [CrossRef] [PubMed]
43. Urashima, M.; Segawa, T.; Okazaki, M.; Kurihara, M.; Wada, Y.; Ida, H. Randomized trial of vitamin D supplementation to prevent seasonal influenza A in schoolchildren. *Am. J. Clin. Nutr.* **2010**, *91*, 1255–1260. [CrossRef] [PubMed]
44. Majak, P.; Olszowiec-Chlebna, M.; Smejda, K.; Stelmach, I. Vitamin D supplementation in children may prevent asthma exacerbation triggered by acute respiratory infection. *J. Allergy Clin. Immunol.* **2011**, *127*, 1294–1296. [CrossRef] [PubMed]
45. Camargo, C.A., Jr.; Clark, S.; Kaplan, M.S.; Lieberman, P.; Wood, R.A. Regional differences in EpiPen prescriptions in the United States: The potential role of vitamin D. *J. Allergy Clin. Immunol.* **2007**, *120*, 131–136. [CrossRef] [PubMed]
46. Camargo, C.A., Jr.; Ingham, T.; Wickens, K.; Thadhani, R.; Silvers, K.M.; Epton, M.J.; Town, G.I.; Pattemore, P.K.; Espinola, J.A.; Crane, J. Cord-blood 25-hydroxyvitamin D levels and risk of respiratory infection, wheezing, and asthma. *Pediatrics* **2011**, *127*, e180–e187. [CrossRef] [PubMed]
47. Heine, G.; Niesner, U.; Chang, H.D.; Steinmeyer, A.; Zugel, U.; Zuberbier, T.; Radbruch, A. Worm, M. 1,25-dihydroxyvitamin D(3) promotes IL-10 production in human B cells. *Eur. J. Immunol.* **2008**, *38*, 2210–2218. [CrossRef]
48. Drozdenko, G.; Scheel, T.; Heine, G.; Baumgrass, R.; Worm, M. Impaired T cell activation and cytokine production by calcitriol-primed human B cells. *Clin. Exp. Immunol.* **2014**, *178*, 364–372. [CrossRef]
49. Wittke, A.; Weaver, V.; Mahon, B.D.; August, A.; Cantorna, M.T. Vitamin D receptor deficient mice fail to develop experimental allergic asthma. *J. Immunol.* **2004**, *173*, 3432–3436. [CrossRef] [PubMed]
50. Hartmann, B.; Heine, G.; Babina, M.; Steinmeyer, A.; Zugel, U.; Radbruch, A.; Worm, M. Targeting the vitamin D receptor inhibits the B cell-dependent allergic immune response. *Allergy* **2011**, *66*, 540–548. [CrossRef]
51. Széles, L.; Keresztes, G.; Töröcsik, D.; Balajthy, Z.; Krenács, L.; Póliska, S.; Steinmeyer, A.; Zuegel, U.; Pruenster, M.; Rot, A.; et al. 1,25-dihydroxyvitamin D3 is an autonomous regulator of the transcriptional changes leading to a tolerogenic dendritic cell phenotype. *J. Immunol.* **2009**, *182*, 2074–2083. [CrossRef] [PubMed]
52. Unger, W.W.; Laban, S.; Kleijwegt, F.S.; van der Slik, A.R.; Roep, B.O. Induction of Treg by monocyte-derived DC modulated by vitamin D3 or dexamethasone: Differential role for PD-L1. *Eur. J. Immunol.* **2009**, *39*, 3147–3159. [CrossRef] [PubMed]

53. Bakdash, G.; van Capel, T.M.; Mason, L.M.; Kapsenberg, M.L.; de Jong, E.C. Vitamin D3 metabolite calcidiol primes human dendritic cells to promote the development of immunomodulatory IL-10-producing T cells. *Vaccine* **2014**, *32*, 6294–6302. [CrossRef] [PubMed]
54. Almerighi, C.; Sinistro, A.; Cavazza, A.; Ciaprini, C.; Rocchi, G.; Bergamini, A. 1α,25-dihydroxyvitamin D3 inhibits CD40L-induced pro-inflammatory and immunomodulatory activity in human monocytes. *Cytokine* **2009**, *45*, 190–197. [CrossRef]
55. Jeffery, L.E.; Burke, F.; Mura, M.; Zheng, Y.; Qureshi, O.S.; Hewison, M.; Walker, L.S.; Lammas, D.A.; Raza, K.; Sansom, D.M. 1,25-Dihydroxyvitamin D3 and IL-2 combine to inhibit T cell production of inflammatory cytokines and promote development of regulatory T cells expressing CTLA-4 and FoxP3. *J. Immunol.* **2009**, *183*, 5458–5467. [CrossRef]
56. Murdaca, G.; Greco, M.; Tonacci, A.; Negrini, S.; Borro, M.; Puppo, F.; Gangemi, S. IL-33/IL-31 Axis in Immune-Mediated and Allergic Diseases. *Int. J. Mol. Sci.* **2019**, *20*, 5856. [CrossRef] [PubMed]
57. Fischer, K.D.; Agrawal, D.K. Hematopoietic Stem and Progenitor Cells in Inflammation and Allergy. *Front. Immunol.* **2013**, *4*, 428. [CrossRef] [PubMed]
58. Uasuf, C.G.; Sano, C.D.; Gangemi, S.; Albeggiani, G.; Cigna, D.; Dino, P.; Brusca, I.; Gjomarkaj, M.; Pace, E. IL-33/s-ST2 ratio, systemic symptoms, and basophil activation in Pru p 3-sensitized allergic patients. *Inflamm. Res.* **2018**, *67*, 671–679. [CrossRef] [PubMed]
59. Ginaldi, L.; De Martinis, M.; Ciccarelli, F.; Saitta, S.; Imbesi, S.; Mannucci, C.; Gangemi, S. Increased levels of interleukin 31 (IL-31) in osteoporosis. *BMC Immunol.* **2015**, *16*, 60. [CrossRef] [PubMed]
60. Tsai, M.; Grimbaldeston, M.; Galli, S.J. Mast cells and immunoregulation/immunomodulation. *Adv. Exp. Med. Biol.* **2011**, *716*, 186–211. [PubMed]
61. Voehringer, D. Protective and pathological roles of mast cells and basophils. *Nat. Rev. Immunol.* **2013**, *13*, 362–375. [CrossRef] [PubMed]
62. Ribatti, D.; Crivellato, E. The role of mast cell in tissue morphogenesis. Thymus, duodenum, and mammary gland as examples. *Exp. Cell Res.* **2016**, *341*, 105–109. [CrossRef] [PubMed]
63. Galli, S.J.; Kalesnikoff, J.; Grimbaldeston, M.A.; Piliponsky, A.M.; Williams, C.M.; Tsai, M. Mast cells as "tunable" effector and immunoregulatory cells: Recent advances. *Ann. Rev. Immunol.* **2005**, *23*, 749–786. [CrossRef] [PubMed]
64. Madjene, L.C.; Danelli, L.; Dahdah, A.; Vibhushan, S.; Bex-Coudrat, J.; Pacreau, E.; Vaugier, C.; Claver, J.; Rolas, L.; Pons, M.; et al. Mast cell chymase protects against acute ischemic kidney injury by limiting neutrophil hyperactivation and recruitment. *Kidney Int.* **2020**, *97*, 516–527. [CrossRef]
65. Katsanos, G.S.; Anogeianaki, A.; Orso, C.; Tete, S.; Salini, V.; Antinolfi, P.L.; Sabatino, G. Mast cells and chemokines. *J. Biol. Regul. Homeost. Agents* **2008**, *22*, 145–151. [PubMed]
66. Lieberman, P.; Garvey, L.H. Mast cells and anaphylaxis. *Curr. Allergy Asthma Rep.* **2016**, *16*, 20. [CrossRef]
67. Finkelman, F.D.; Khodoun, M.V.; Strait, R. Human IgE-independent systemic anaphylaxis. *J. Allergy Clin. Immunol.* **2016**, *137*, 1674–1680. [CrossRef] [PubMed]
68. Afrin, L.B. Mast cell activation disease and the modern epidemic of chronic inflammatory disease. *Transl. Res.* **2016**, *174*, 33–59. [CrossRef]
69. Wernersson, S.; Pejler, G. Mast cell secretory granules: Armed for battle. *Nat. Rev. Immunol.* **2014**, *14*, 478–494. [CrossRef]
70. Liu, Z.Q.; Li, X.X.; Qiu, S.Q.; Yu, Y.; Li, M.G.; Yang, L.T.; Li, L.J.; Wang, S.; Zheng, P.Y.; Liu, Z.G.; et al. Vitamin D contributes to mast cell stabilization. *Allergy* **2017**, *72*, 1184–1192. [CrossRef] [PubMed]
71. Biggs, L.; Yu, C.; Fedoric, B.; Lopez, A.F.; Galli, S.J.; Grimbaldeston, M.A. Evidence that vitamin D(3) promotes mast cell-dependent reduction of chronic UVB-induced skin pathology in mice. *J. Exp. Med.* **2010**, *207*, 455–463. [CrossRef]
72. Asero, R.; Ferrucci, S.; Casazza, G.; Marzano, A.V.; Cugno, M. Total IgE and atopic status in patients with severe chronic spontaneous urticaria unresponsive to omalizumab treatment. *Allergy* **2019**, *74*, 1561–1563. [CrossRef]
73. Lakin, E.; Church, M.K.; Maurer, M.; Schmetzer, O. On the Lipophilic Nature of Autoreactive IgE in Chronic Spontaneous Urticaria. *Theranostics* **2019**, *9*, 829–836. [CrossRef] [PubMed]
74. Rivero-Yeverino, D.; López-García, A.I.; Caballero-López, C.G.; Ríos-López, J.J.; Papaqui-Tapia, J.S.; Ortega-Jordá Rodríguez, E.E. Álvarez-Rivera, A.; Ruiz-Sánchez, D.M.; Flores-Gonzaga, E. Vitamin D and respiratory allergy: State of the art. *Rev. Alergy Mex* **2022**, *69* (Suppl. S1), s46–s54. [CrossRef]
75. He, L.; Yi, W.; Huang, X.; Long, H.; Lu, Q. Chronic Urticaria: Advances in Understanding of the Disease and Clinical Management. *Clin. Rev. Allergy Immunol.* **2021**, *61*, 424–448. [CrossRef] [PubMed]
76. Redegeld, F.A.; Yu, Y.; Kumari, S.; Charles, N.; Blank, U. Non-IgE mediated mast cell activation. *Immunol. Rev.* **2018**, *282*, 87–113. [CrossRef] [PubMed]
77. Metz, M.; Lammel, V.; Gibbs, B.F.; Maurer, M. Inflammatory murine skin responses to UV-B light are partially dependent on endothelin-1 and mast cells. *Am. J. Pathol.* **2006**, *169*, 815–822. [CrossRef] [PubMed]
78. Galli, S.J.; Grimbaldeston, M.; Tsai, T. Immunomodulatory mast cells: Negative, as well as positive, regulators of immunity. *Nat Rev. Immunol.* **2008**, *8*, 478–486. [CrossRef] [PubMed]
79. Hart, P.H.; Townley, S.L.; Grimbaldeston, M.A.; Khalil, Z.; Finlay Jones, J.J. Mast cells, neuropeptides, histamine, and prostaglandins in UV-induced systemic immunosuppression. *Methods* **2002**, *28*, 79–89. [CrossRef]
80. Khalil, Z.; Townley, S.L.; Grimbaldeston, M.A.; Finlay-Jones, J.J.; Hart, P.H. cis-Urocanic acid stimulates neuropeptide release from peripheral sensory nerves. *J. Investig. Dermatol.* **2002**, *117*, 886–891. [CrossRef]

81. Yip, K.H.; Kolesnikoff, N.; Yu, C.; Hauschild, N.; Taing, H.; Biggs, L.; Goltzman, D.; Gregory, P.A.; Anderson, P.H.; Samuel, M.S.; et al. Mechanisms of vitamin D_3 metabolite repression of IgE-dependent mast cell activation. *J. Allergy Clin. Immunol.* **2014**, *133*, 1356–1364.e14. [CrossRef]
82. Lindholm, R.; Lindholm, S.; Liukko, P. Fracture healing and mast cells. I. The periosteal callus in rats. *Acta Orthop. Scand.* **1967**, *38*, 115–122. [CrossRef] [PubMed]
83. Banovac, K.; Renfree, K.; Makowski, A.L.; Latta, L.L.; Altman, R.D. Fracture healing and mast cells. *J. Orthop. Traum.* **1995**, *9*, 482–490. [CrossRef] [PubMed]
84. Antebi, B.; Zhang, L.; Sheyn, D.; Pelled, G.; Zhang, X.; Gazit, Z.; Schwarz, E.M.; Gazit, D. Controlling Arteriogenesis and Mast Cells Are Central to Bioengineering Solutions for Critical Bone Defect Repair Using Allografts. *Bioengineering* **2016**, *3*, 6. [CrossRef]
85. Zhang, L.; Wang, T.; Chang, M.; Kaiser, C.; Kim, J.D.; Wu, T.; Cao, X.; Zhang, X.; Schwarz, E.M. Teriparatide Treatment Improves Bone Defect Healing Via Anabolic Effects on New Bone Formation and Non-Anabolic Effects on Inhibition of Mast Cells in a Murine Cranial Window Model. *J. Bone Miner. Res.* **2017**, *32*, 1870–1883. [CrossRef] [PubMed]
86. Silveira, S.R.; Hadler, W.A. A histochemical investigation on the percutaneous absorption of vitamin D synthesized into the mammal epidermis. *Acta Histochem.* **1985**, *77*, 11–18. [CrossRef]
87. Metcalfe, D.D. Mast cells and mastocytosis. *Blood* **2008**, *112*, 946–956. [CrossRef] [PubMed]
88. Arock, M.; Valent, P. Pathogenesis, classification and treatment of mastocytosis: State of the art in 2010 and future perspectives. *Exp. Rev. Hematol.* **2010**, *3*, 497–516. [CrossRef]
89. Pardanani, A. Systemic mastocytosis in adults: 2019 update on diagnosis, risk stratification and management. *Am. J. Hematol.* **2019**, *94*, 363–377. [CrossRef] [PubMed]
90. Valent, P.; Escribano, L.; Broesby-Olsen, S.; Hartmann, K.; Grattan, C.; Brockow, K.; Niedoszytko, M.; Nedoszytko, B.; Oude Elberink, J.N.; Kristensen, T.; et al. Proposed diagnostic algorithm for patients with suspected mastocytosis: A proposal of the European Competence Network on Mastocytosis. *Allergy* **2014**, *69*, 1267–1274. [CrossRef] [PubMed]
91. Monnier, J.; Georgin-Lavialle, S.; Canioni, D.; Lhermitte, L.; Soussan, M.; Arock, M.; Bruneau, J.; Dubreuil, P.; Bodemer, C.; Chandesris, M.O.; et al. Mast cell sarcoma: New cases and literature review. *Oncotarget* **2016**, *7*, 66299–66309. [CrossRef]
92. Kanis, J.A. Assessment of fracture risk and its application to screening for postmenopausal osteoporosis: Synopsis of a WHO report. WHO study Group. *Osteop. Int.* **1994**, *4*, 368–381. [CrossRef] [PubMed]
93. Rossini, M.; Zanotti, R.; Viapiana, O.; Tripi, G.; Orsolini, G.; Idolazzi, L.; Bonadonna, P.; Schena, D.; Escribano, L.; Adami, S.; et al. Bone involvement and osteoporosis in mastocytosis. *Immunol. Allergy Clin. N. Am.* **2014**, *34*, 383–396. [CrossRef] [PubMed]
94. Rossini, M.; Zanotti, R.; Orsolini, G.; Tripi, G.; Viapiana, O.; Idolazzi, L.; Zamo, A.; Bonadonna, P.; Kunnathully, V.; Adami, S.; et al. Prevalence, pathogenesis, and treatment options for mastocytosis-related osteoporosis. *Osteop. Int.* **2016**, *27*, 2411–2421. [CrossRef] [PubMed]
95. Lewiecki, E.M.; Gordon, C.M.; Baim, S.; Leonard, M.B.; Bishop, N.J.; Bianchi, M.L. International society for clinical densitometry 2007 adult and pediatric official positions. *Bone* **2008**, *43*, 1115–1121. [CrossRef] [PubMed]
96. Rossini, M.; Zanotti, R.; Bonadonna, P.; Artuso, A.; Caruso, B.; Schena, D.; Vecchiato, D.; Bonifacio, M.; Viapiana, O.; Gatti, D.; et al. Bone mineral density, bone turnover markers and fractures in patients with indolent systemic mastocytosis. *Bone* **2011**, *49*, 880–885. [CrossRef] [PubMed]
97. Valent, P.; Akin, C.; Gleixner, K.V.; Sperr, W.R.; Reiter, A.; Arock, M.; Triggiani, M. Multidisciplinary Challenges in Mastocytosis and How to Address with Personalized Medicine Approaches. *Int. J. Mol. Sci.* **2019**, *20*, 2976. [CrossRef]
98. Fallon, M.D.; Whyte, M.P.; Teitelbaum, S.L. Systemic mastocytosis associated with generalized osteopenia. Histopathological characterization of the skeletal lesion using undecalcified bone from two patients. *Hum. Pathol.* **1981**, *12*, 813–820. [CrossRef]
99. Manara, M.; Varenna, M.; Cantoni, S.; Parafioriti, A.; Gallazzi, M.B.; Sinigaglia, L. Osteoporosis with vertebral fractures in young males, due to bone marrow mastocytosis: A report of two cases. *Clin. Exp. Rheumatol.* **2010**, *28*, 97–100.
100. Orsolini, G.; Viapiana, O.; Rossini, M.; Bonifacio, M.; Zanotti, R. Bone Disease in Mastocytosis. *Immunol Allergy Clin. N. Am.* **2018**, *38*, 443–454. [CrossRef]
101. Artuso, A.; Caimmi, C.; Tripi, G.; Viapiana, O.; Bonifacio, M.; Idolazzi, L.; Gavioli, I.; Gate, D.; Zanotti, R.; Rossini, M. Longitudinal Evaluation of Bone Mineral Density and Bone Metabolism Markers in Patients with Indolent Systemic Mastocytosis Without Osteoporosis. *Calcif. Tissue Int.* **2017**, *100*, 40–46. [CrossRef] [PubMed]
102. Cohen, D.; Reif, J.S.; Brodey, R.S.; Keiser, H. Epidemiological analysis of the most prevalent sites and types of canine neoplasia observed in a veterinary hospital. *Cancer Res.* **1974**, *34*, 2859–2868. [PubMed]
103. Wakshlag, J.J.; Rassnick, K.M.; Malone, E.K.; Struble, A.M.; Vachhani, P.; Trump, D.L.; Tian, L. Cross-sectional study to investigate the association between vitamin D status and cutaneous mast cell tumours in Labrador retrievers. *Br. J. Nutr.* **2011**, *106* (Suppl. S1), S60–S63. [CrossRef] [PubMed]
104. Rassnick, K.M.; Muindi, J.R.; Johnson, C.S.; Balkman, C.E.; Ramnath, N.; Yu, W.D.; Engler, K.L.; Page, R.L.; Trump, D.L. In vitro and in vivo evaluation of combined calcitriol and cisplatin in dogs with spontaneously occurring tumors. *Cancer Chemother. Pharmacol.* **2008**, *62*, 881–981. [CrossRef] [PubMed]
105. Russell, D.S.; Rassnick, K.M.; Erb, H.N.; Vaughan, M.M.; McDonough, S.P. An immunohistochemical study of vitamin D receptor expression in canine cutaneous mast cell tumours. *J. Comp. Pathol.* **2010**, *143*, 223–226. [CrossRef] [PubMed]

106. Guillaume, N.; Desoutter, J.; Chandesris, O.; Merlusca, L.; Henry, I.; Georgin-Lavialle, S.; Barete, S.; Hirsch, I.; Bouredji, D.; Roye B.; et al. Bone complications of mastocytosis: A link between clinical and biological characteristics. *Am. J. Med.* **2013**, *126*, 7 [CrossRef]
107. Orsolini, G.; Gavioli, I.; Tripi, G.; Viapiana, O.; Gatti, D.; Idolazzi, L.; Zanotti, R.; Rossini, M. Denosumab for the Treatment o Mastocytosis-Related Osteoporosis: A Case Series. *Calcif. Tissue Int.* **2017**, *100*, 595–598. [CrossRef] [PubMed]
108. Roodman, G.D. Biology of osteoclast activation in cancer. *J. Clin. Oncol.* **2001**, *19*, 3562–3571. [CrossRef] [PubMed]
109. King, J.J.; Crawford, E.A.; Iwenofu, O.H.; Fox, E.J. Case report: Pathologic long bone fracture in a patient with systemi mastocytosis. *Clin. Orthop. Relat. Res.* **2007**, *459*, 263–269. [CrossRef]
110. Garla, V.V.; Chaudhary, K.U.Q.; Yaqub, A. Systemic mastocytosis: A rare cause of osteoporosis. *Pan Afr. Med. J.* **2019**, *32*, 16 [CrossRef]
111. Rabenhorst, A.; Christopeit, B.; Leja, S.; Gerbaulet, A.; Kleiner, S.; Forster, A.; Raap, U.; Wickenhauser, C.; Hartmann, K. Serur levels of bone cytokines are increased in indolent systemic mastocytosis associated with osteopenia or osteoporosis. *J. Allerg Clin. Immunol.* **2013**, *132*, 1234–1237. [CrossRef] [PubMed]
112. Renke, J.; Kędzierska-Mieszkowska, S.; Lange, M.; Nedoszytko, B.; Wasilewska, E.; Liberek, A.; Renke, M.; Niedoszytko, M Witkowski, J.; Skórko-Glonek, J.; et al. Mast cells in mastocytosis and allergy—Important player in metabolic and immunologica homeostasis. *Adv. Med. Sci.* **2019**, *64*, 124–130. [CrossRef] [PubMed]
113. Bossi, F.; Frossi, B.; Radillo, O.; Cugno, M.; Tedeschi, A.; Riboldi, P.; Asero, R.; Tedesco, F.; Pucillo, C. Mast cells are criticall involved in serum-mediated vascular leakage in chronic urticaria beyond high-affinity IgE receptor stimulation. *Allergy* **2011**, *6* 1538–1545. [CrossRef] [PubMed]
114. Zhao, J.W.; Ping, J.D.; Wang, Y.F.; Liu, X.N.; Li, N.; Hu, Z.L.; Ming, L. Vitamin D suppress the production of vascular endotheli growth factor in mast cell by inhibiting PI3K/Akt/p38 MAPK/HIF-1α pathway in chronic spontaneous urticaria. *Clin. Immun* **2020**, *215*, 108444. [CrossRef] [PubMed]
115. Bader-Meunier, B.; Livideanu, C.B.; Larroche, C.; Durieu, I.; Artru, L.; Beucher, A.; Cormier, G.; Cornec, D.; DeLarco, M.; Dubos J.-J.; et al. Association of mastocytosis with inflammatory joint diseases: A series of 31 patients. *Semin. Arthritis Rheum.* **2014**, *4* 362–365. [CrossRef]
116. Metcalfe, D.D. Mastocytosis. In *Middleton's Allergy: Principles and Practice*, 9th ed.; Wesley, B.A., Ed.; Elsevier Inc.: Amsterdan The Netherlands, 2020; Volume 74, pp. 1216–1227.
117. Ch'ng, S.; Wallis, R.A.; Yuan, L.; Davis, P.F.; Tan, S.T. Mast cells and cutaneous malignancies. *Mod. Pathol.* **2006**, *19*, 149–15 [CrossRef]
118. Harvima, I.T.; Nilsson, G. Mast cells as regulators of skin inflammation and immunity. *Acta Derm. Venereol.* **2011**, *91*, 644–65 [CrossRef] [PubMed]
119. Murphy, G.M. Ultraviolet radiation and immunosuppression. *Br. J. Dermatol.* **2009**, *161*, 90–95. [CrossRef] [PubMed]
120. Yu, C.; Fedoric, B.; Anderson, P.H.; Lopez, A.F.; Grimbaldeston, M.A. Vitamin D3 signalling to mast cells: A new regulatory axi *Int. J. Biochem. Cell Biol.* **2011**, *43*, 41–46. [CrossRef]
121. Babina, M.; Krautheim, M.; Grützkau, A.; Henz, B.M. Human leukemic (HMC-1) mast cells are responsive to 1,2 dihydroxyvitamin D3: Selective promotion of ICAM-3 expression and constitutive presence of vitamin D3 receptor. *Biocher Biophys. Res. Commun.* **2000**, *273*, 1104–1110. [CrossRef]
122. Kaukinen, A.; Pelkonen, J.; Harvima, I.T. Mast cells express CYP27A1 and CYP27B1 in epithelial skin cancers and psoriasis. *Eur. Dermatol.* **2015**, *25*, 548–555. [CrossRef] [PubMed]
123. Jones, G.; Prosser, D.E.; Kaufmann, M. Cytochrome P450-mediated metabolism of vitamin D. *J. Lipid Res.* **2014**, *55*, 13–3 [CrossRef] [PubMed]
124. Lou, Y.R.; Molnár, F.; Peräkylä, M.; Qiao, S.; Kalueff, A.V.; St-Arnaud, R.; Carlberg, C.; Tuohimaa, P. 25-Hydroxyvitamin D(3) is a agonistic vitamin D receptor ligand. *J. Steroid Biochem. Mol. Biol.* **2010**, *118*, 162–170. [CrossRef]
125. Kaukinen, A.; Siiskonen, H.; Pelkonen, J.; Harvima, I.T. Immunoreactivity to CYP24A1, but not vitamin D receptor, is increased i mast cells of keratinocyte skin cancers. *Eur. J. Dermatol.* **2017**, *27*, 590–598. [CrossRef] [PubMed]
126. Hypponen, E.; Berry, D.J.; Wjst, M.; Power, C. Serum 25-hydroxyvitamin D and IgE-a significant but nonlinear relationshi *Allergy* **2009**, *64*, 613–620. [CrossRef]
127. Rothers, J.; Wright, A.L.; Stern, D.A.; Halonen, M.; Camargo, C.A., Jr. Cord blood 25-hydroxyvitamin D levels are associate with aeroallergen sensitization in children from Tucson, Arizona. *J. Allergy Clin. Immunol.* **2011**, *128*, 1093–1099.e5. [CrossRe [PubMed]
128. Cairncross, C.; Grant, C.; Stonehouse, W.; Conlon, C.; McDonald, B.; Houghton, L.; Eyles, D.; Camargo, C.A.; Coad, J.; von Hur P. The Relationship between Vitamin D Status and Allergic Diseases in New Zealand Preschool Children. *Nutrients* **2016**, *8*, 32 [CrossRef] [PubMed]
129. Li, M.; Hener, P.; Zhang, Z.; Kato, S.; Metzger, D.; Chambon, P. Topical vitamin D3 and low-calcemic analogs induce thym stromal lymphopoietin in mouse keratinocytes and trigger an atopic dermatitis. *Proc. Natl. Acad. Sci. USA* **2006**, *103*, 11736–1174 [CrossRef] [PubMed]

biomedicines

MDPI

Systematic Review

Systematic Review of the Treatment of Persistent Hyperparathyroidism Following Kidney Transplantation

Miłosz Miedziaszczyk [1,*,†], Katarzyna Lacka [2,†], Oskar Tomczak [3], Aleksander Bajon [4], Marta Primke [5] and Ilona Idasiak-Piechocka [1]

1 Department of Nephrology, Transplantology and Internal Medicine, Poznan University of Medical Sciences, 60-355 Poznan, Poland
2 Department of Endocrinology, Metabolism and Internal Medicine, Poznan University of Medical Sciences, 60-355 Poznan, Poland
3 1st Department of Anesthesiology and Intensive Therapy, Poznan University of Medical Sciences, 61-848 Poznan, Poland
4 Student Scientific Section of Department of Nephrology, Transplantology and Internal Medicine, Poznan University of Medical Sciences, 60-355 Poznan, Poland
5 Student's Scientific Section of Department of Endocrinology, Metabolism and Internal Medicine, Poznan University of Medical Sciences, 60-355 Poznan, Poland
* Correspondence: m.miedziaszczyk@wp.pl
† These authors contributed equally to this manuscript.

Abstract: Chronic kidney disease–mineral and bone disorder is one of the complications associated with chronic kidney disease. About 10–50% of patients following kidney transplantation have persistent hyperparathyroidism. Hypercalcaemic hyperparathyroidism has a negative impact on the kidney transplant outcome; therefore, it requires treatment. The data regarding the treatment of persistent hyperparathyroidism provided in scientific publications are divergent and contradictory. Therefore, the aim of our systematic review was to evaluate the efficacy of persistent hyperparathyroidism treatment in patients following kidney transplantation. The Cochrane, PubMed, and Scopus databases were browsed independently by two authors. The search strategy included controlled vocabulary and keywords. The effectiveness of calcitriol, paricalcitol, cinacalcet, and parathyroidectomy was compared and analysed. The mean calcium and parathormone (PTH) concentrations per patient in the group of paricalcitol increased by 1.27% and decreased by 35.14% (n = 248); in the group of cinacalcet decreased by 12.09% and 32.16% (n = 368); and in the group of parathyroidectomy decreased by 19.06% and 86.49% (n = 15) at the end of the study compared to the baseline (n = 244, n = 342 and n = 15), respectively. Paricalcitol, cinacalcet, and parathyroidectomy decreased the intact PTH level. Cinacalcet and parathyroidectomy lowered calcium levels in renal transplant patients with hypercalcaemia. Conversely, paricalcitol increased the serum calcium concentration. Cinacalcet seems to be a good candidate in the treatment of post-transplant hyperparathyroidism.

Keywords: persistent hyperparathyroidism; kidney transplantation; renal transplantation; treatment; systematic review

1. Introduction

Secondary hyperparathyroidism (SHPT) is a condition in which the parathormone (PTH) level is elevated to compensate for the calcium–phosphorus imbalance. An increase in PTH secretion is triggered by hypocalcaemia, hyperphosphataemia, or a decreased active vitamin D [1]. In pre-kidney transplantation (pre-KTx) patients with advanced chronic kidney disease (CKD), calcitriol (active vitamin D) deficiency is related to the inability to convert vitamin D to its physiologically active form, 1,25-cholecalciferol. Subsequently, it causes a decrease in intestinal absorption of calcium and a low serum calcium level [2]. Furthermore, a decreasing glomerular filtration rate (GFR) in CKD leads to reduced phosphate clearance and hyperphosphataemia [3]. In turn, excess serum phosphorus increases

PTH secretion, as well as results in hypocalcaemia (phosphorus forms complexes with calcium) and stimulates fibroblast growth factor-23 (FGF-23) [4]. SHPT is one of the most common complications associated with CKD. Mineral metabolism imbalances increase the risk of bone fractures, renal osteodystrophy, and vascular calcification. This, in turn, contributes to the development of cardiovascular diseases and results in higher mortality rates among CKD patients [5]. Moreover, prolonged exposure to the aforementioned electrolyte disturbance leads to hyperplasia of parathyroid tissue, which persists even after kidney transplantation [2]. One of the accompanying conditions related to CKD is SHPT, which may remain problematic post-transplantation [6]. The incidence of persistent hyperparathyroidism after kidney transplantation is estimated to be 10–50%; it depends on the age group and increases with age [7–9]. There is no consensus regarding a PTH level which clearly defines the presence of persistent post-transplant hyperparathyroidism. Nevertheless, most clinicians consider a PTH level greater than two times normal (>130 pg/mL) to be consistent with persistent post-transplant hyperparathyroidism (PT-HPT) within 12 months post-transplant [6]. SHPT initially manifests as polyclonal diffuse hyperplasia, which then evolves into multiple monoclonal nodules. Lesion advancement correlates with the increasing weight of the parathyroid gland, although even in low-weight parathyroid glands nodular lesions have been detected [10]. Taniguchi et al. [11] reported that calcium-sensing receptors, vitamin D receptors, as well as apoptosis increased following transplantation with diffuse hyperplasia; therefore, shrinkage of this part is likely to be observed. This is associated with a gradual decrease of PTH levels after transplantation observed in some patients [12,13]. The remaining nodular hyperplastic part of the parathyroid gland does not demonstrate similar receptor changes, thus indicating that they account for most tertiary HPT effects [11]. Studies showed that the most significant independent risk factors of post-transplant HPT include high pre-transplantation levels of PTH and glomerular filtration rate of the kidney transplant. Other, such as long-term duration of dialysis, large maximum parathyroid gland size, high (<2 weeks) Ca level, and high post-transplant (<2 weeks) ALP concentration contribute to the exacerbation of this abnormality. Thus, early control of risk factors could positively affect post-transplant outcomes [12–15]. Additionally, after the transplantation, suboptimal graft function, immunosuppressive drugs, metabolic acidosis, hyperphosphataemia, and low vitamin D levels also may increase PTH secretion and contribute to persistent HPT [16]. Moreover, HPT associated with the deterioration of renal transplant function is also of note, when PTH levels increase along with the deterioration of renal transplant function to maintain normal phosphataemia and normocalcaemia [17]. Persistent hypercalcaemia in post-transplant patients plays a role in the early calcification of renal allografts [18] and progression of artery calcification [19]. In transplantation patients, unlike the dialysis patients, an increased PTH level is accompanied by hypophosphataemia. In fact, these two factors, i.e., high PTH and low phosphate, may affect bone metabolism differently. An elevated PTH level seems to have a protective effect on osteoblasts by means of preserving their survival. In contrast, a decreased phosphorus level has been associated with impaired osteoblastogenesis and early osteoblast apoptosis. Therefore, the abovementioned divergent effects may result in low bone turnover [20]. Despite mineral balance abnormalities, SHPT may lead to other systemic problems. Ivarsson et al. [21] reported that increased PTH was associated with new-onset diabetes mellitus after transplantation, which has a negative impact on both the graft and patient survival.

There is an ongoing debate with regard to the early treatment of persistent post-transplant hyperparathyroidism. On the one hand, some researchers emphasise that this condition often improves with time. However, other studies provide data indicating that improvement in parathyroid function at 3 months post-transplant is unlikely [7]. Another argument is that the withdrawal of some medications, such as cinacalcet, at the time of renal transplantation does not constitute a risk factor for allograft calcifications in the early post-transplantation period [22]. Nevertheless, hypercalcaemic hyperparathyroidism itself has a negative impact on the kidney transplant outcome; therefore, it requires treatment. Furthermore, the first year of the post-transplantation period is essential in terms of bone loss [23]. In addition to its role in bone disease, elevated PTH level is a risk factor of vascular calcification, which leads to an increased post-transplant cardiovascular morbidity and mortality [24].

2. Materials and Methods

The aim of our systematic review was to evaluate the efficacy of persistent hyperparathyroidism treatment in patients following kidney transplantation (KTx).

The search strategy included controlled vocabulary and keywords. The Cochrane, PubMed, and Scopus databases were browsed independently by two authors. The main search criteria were 'persistent hyperparathyroidism' AND 'renal transplantation' or 'kidney transplantation' AND 'cinacalcet' or 'paricalcitol' or 'parathyroidectomy'. Studies published in the past 13 years were considered for this review (Figure 1). The PRISMA Checklist was verified during the draft of the review. A flowchart of the systematic literature search was developed according to PRISMA guidelines. Study protocol has been registered (ID CRD42022369857). The papers were selected depending on their clinical importance and Greenhalgh's evidence hierarchy. Therefore, case reports were considered as having low clinical importance; nevertheless, they were mentioned in our review in order to provide a broader perspective on the topic. Data were collected, arranged, and subsequently the selected information was depicted in Tables 1 and 2. Nevertheless, due to no data on the topic, it was impossible to include the presence of primary hyperparathyroidism as an exclusion criterion. Additionally, we could not provide the number of treated SHPT patients and the number of the remaining HPT cases. Statistical calculations were performed using the direct method of arithmetic mean of grouped data.

Table 1. Selected information with regard to the analysed studies [25–37].

The Active Substance	Dosage	Time Study [Weeks]	Study Group (n) * Completed	Control Group (n) * Completed	Placebo or Comparator
Paricalcitol [27]	1 μg/day and next 2 μg/day	24	22 * 20	21 * 19	other therapy
Paricalcitol [25]	2 μg/day	44	37 * 35	40	standard therapy
Paricalcitol [26]	2–3 μg/day	96	69	-	-
Paricalcitol [28]	1 μg/day	72	58	-	-
Paricalcitol [35]	1 μg/day	12	31	-	-
Paricalcitol [36]	1 μg/day	48	31	31	standard therapy
Cinacalcet [30]	30 mg/day	48	32 * 29	-	-
Cinacalcet [29]	30 mg/day	48	18	-	-
Cinacalcet [31]	30 mg/day	24	193 * 175	-	-
Cinacalcet [32]	30 mg/day	24	47 * no data	-	-
Cinacalcet [33]	30 mg/day	240	6	-	-
Cinacalcet [34]	30 mg/day	52	57 * 52	57 * 52	placebo
Cinacalcet, Parathyreidectomy [37]	30 mg every other day 30 mg/day 60 mg/day	240	15 * 13	15 * 11	parathyreidectomy

* denotes number of patients who completed the study.

Table 2. Changes in the serum concentration of PTH, calcium, phosphorus, and the value of albumin, or information regarding the corrected serum calcium concentration based on the albumin level [25–37].

The Active Substance	PTH before the Study [pg/mL]	PTH after the Study [pg/mL]	Phosphorus Serum before [mmol/L]	Phosphorus Serum after [mmol/L]	Calcium Serum before [mmol/L]	Calcium Serum after [mmol/L]	Albumin (Serum Calcium Concentrations Were Corrected for Albumin Level)
Paricalcitol [27]	115.60	63.25	1.06	1.10	2.38	2.40	no data
Paricalcitol [25]	105.62	93.36	0.93	0.95	2.37	2.39	no data
Paricalcitol [26]	288	193	1.10	1.16	2.42	2.40	no data
Paricalcitol [28]	333	181	1.10	1.13	2.32	2.40	no data
Paricalcitol [35]	216	167	1.05	1.04	2.41	2.43	no data
Paricalcitol [36]	100.00	59.70	1.00	1.01	2.33	2.36	serum calcium concentrations were corrected for the albumin level
Cinacalcet [30]	364	145	1.03	1.07	2.37	2.30	no data
Cinacalcet [29]	242.04	145.62	0.65	0.94	2.78	2.48	no data
Cinacalcet [31]	235	181	0.87	0.97	2.77	2.52	no data
Cinacalcet [32]	310.72	238.39	0.93	0.96	2.69	2.48	serum calcium concentrations were corrected for the albumin level
Cinacalcet [33]	260	237	0.90	1.01	2.74	2.56	no data
Cinacalcet [34]	327.7	169.0	0.86	1.06	2.81	2.39	serum calcium concentrations were corrected for the albumin level
Cinacalcet [37]	235.75	198.03	0.92	0.96	2.72	2.43	serum calcium concentrations were corrected for the albumin level
Parathyreidectomy [37]	348.91	47.15	0.93	1.30	2.78	2.25	

Studies [37,38] represent the same group of patients; the difference is in the duration of the study.

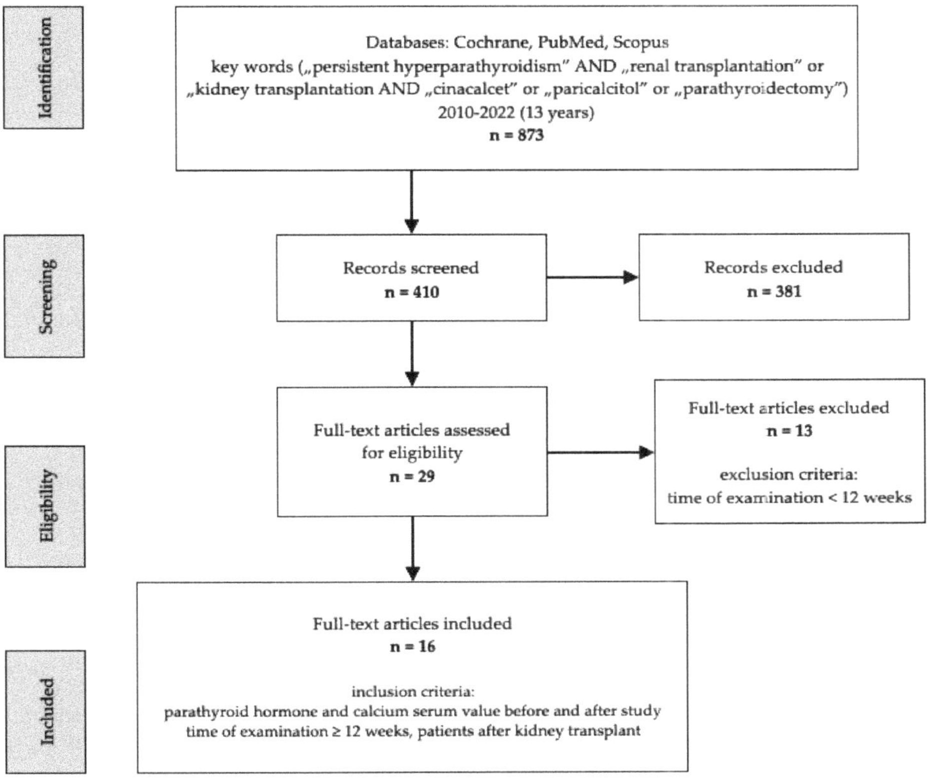

Figure 1. Stages of including articles in the review, PRISMA flowchart.

3. Results

The mean calcium concentration per patient in the paricalcitol efficacy studies increased by 1.27% (n = 248) at the end of the study compared to the baseline (n = 244). The mean intact PTH (iPTH) concentration per patient decreased by 35.14% (n = 248) at the end of the study compared to the baseline (n = 244). The mean calcium concentration per patient in the cinacalcet efficacy studies decreased by 12.09% (n = 368) at the end of the study (n = 342). The mean iPTH concentration per patient decreased by 32.16% (n = 368) at the end of the study compared to the baseline (n = 342). The mean calcium concentration per patient in the parathyroidectomy efficacy studies decreased by 19.06% (n = 15) at the end of the study compared to the baseline (n = 15). The mean PTH concentration per patient decreased by 86.49% (n = 15) at the end of the study compared to the baseline (n = 15).

3.1. *Calcitriol*

Calcitriol is an active form of cholecalciferol (vitamin D3). The synthesis of calcitriol requires two hydroxylations. The first takes place in the liver and results in the production of 25-hydroxycholecalciferol (25-hydroxyvitamin D3, calcifediol, 25(OH)D3) from cholecalciferol. Thereafter, due to the action of 1α-hydroxylase in the kidneys, 1α,25-dihydroxyvitamin D3 (calcitriol, 1,25(OH)2D3) is synthesised. Factors which stimulate this process include parathyroid hormone (PTH), low calcium, and low phosphate levels. Conversely, high levels of fibroblast growth factor 23 (FGF23) and phosphate cause a reduction in the synthesis of calcitriol. An active form of vitamin D3 (calcitriol) binds to vitamin D receptor (VDR), which leads to the stimulation of renal and intestinal absorption of calcium and phosphate, as well as a downregulation of PTH synthesis. Additionally, calcitriol has a direct impact on osteoclasts and osteoblasts; it also modulates immune and inflammatory response, as well as renin production [39,40].

Six months after a renal transplant, as a result of lower FGF23 levels, levels of serum calcitriol increase significantly [41,42]. However, immunosuppressive therapy contributes to a decrease in 1,25(OH)2D3. Moreover, glucocorticoids intensify vitamin D catabolism and promote an increase of PTH levels. Among kidney transplant recipients, reduced 25(OH)D levels (ercalcifediol, 25(OH)D2 and 25(OH)D3), as well as a deranged metabolism of vitamin D3 are highly prevalent, which contributes to the presence of chronic kidney disease–mineral and bone disorder (CKD-MBD). Precise mechanisms of alterations in calcitriol metabolism are still insufficiently understood [39]. Several trials considered calcitriol therapy among patients with persistent secondary hyperparathyroidism (SHPT) following renal transplant. According to Tillmann et al. [43], the application of vitamin D therapy, in this clinical situation, is limited. Patients with persistent SHPT frequently suffer from hypercalcaemia, while SHPT (prior to the renal transplant) co-occurs with normocalcaemia or hypocalcaemia. Conversely, calcitriol supplementation reduces PTH levels [44,45] and protects against loss of bone mineral density [40,45]. In fact, Wada et al. [45] observed that a combination of denosumab and calcitriol enabled effective treatment of a patient with persistent SHPT and a massive bone loss. Denosumab in monotherapy resulted in hypocalcaemia, which was effectively improved by calcitriol supplementation. Nonetheless, further research on this treatment method should be conducted. Vitamin D also exhibits a protective impact on the cardiovascular system, and may play a role in the treatment of proteinuria, which is particularly important for patients with CKD [39]. Another application of an active form of vitamin D, combined with calcium supplementation, is in patients following parathyroidectomy, as a form of treatment for persistent SHPT. The studies indicate that it prevents hypocalcaemia, caused by the hungry bone syndrome, which may appear after resection [42]. Nevertheless, due to the scarcity of large clinical trials, an appropriate dose of calcitriol and a precise mode of action remain unknown. There is a great need for further investigation and confirmation of its application in the treatment of persistent SHPT [44].

3.2. Paricalcitiol

Paricalcitiol (19-nor-1,25-dihydroxyvitamin D2) is a synthetic vitamin D analogue which binds to the vitamin D receptors in a similar manner as 1,25-dihydroxyvitamin D3 (calcitriol). Therefore, by affecting the retinoid X receptor, it regulates the transcriptional activity of vitamin D-responsive genes. The crucial result of that action is a decrease in the secretion of parathyroid hormone (PTH) in a dose-dependent manner, which suppresses parathyroid hyperplasia [46]. According to the available sources, the paricalcitol decreasing effect on PTH is undoubted [25–28,47–50]. However, information about the spectrum of action of this medication may differ depending on the specific source. Žilinská et al. [43] reported an increased calcaemia and calciuria, accompanied by diminishing phosphorus levels after paricalcitol treatment. Nevertheless, no case of hypercalcaemia was recorded, and the calcium plasma levels were within the reference range. On the other hand, other studies did not demonstrate any significant fluctuation in calcium blood levels [26,49]. However, bone density measurements and radiographic evaluation following paricalcitol administration showed significant improvement in bone condition [27,47,49], whereas serum bone-specific alkaline phosphatase and osteocalcin decreased as compared to baseline [25,27].

The abovementioned paricalcitol effect on calcium–phosphorus balance and bone turnover may be the result of changes on the genetic level, as well as on inner-cell signalling pathways [48,50]. Therefore, KLOTHO was investigated. This multifunctional protein functions as a co-receptor for fibroblast growth factor-23 (FGF23) to modulate FGF23 signal transduction. The putative physiologic role of FGF23 is to act as a counter-regulatory phosphaturic hormone to maintain phosphate homeostasis in response to vitamin D [50].

Donate-Correa et al. [48] investigated the ratio of the unmethylated/methylated klotho promoter DNA, which did not change in controls, although it was increased by 177% in paricalcitol-treated subjects. The increase in this ratio was independently associated with the change in messenger RNA expression levels, which showed a significant and direct correlation with an elevation of serum klotho concentrations. Subsequently, a significant increase in the serum levels of FGF-23 after paricalcitol administration was detected. Another pleiotropic effect of paricalcitol worth bearing in mind is a significant reduction of C-reactive protein (CRP) serum concentrations. CRP is a best-characterized biomarker of inflammation; thus, reducing its level is a sign of alleviating fibrosis and other inflammation-associated adverse effects on the graft [25,28,48,49]. The majority of tests approved the anti-proteinuric effect of paricalcitol; hence, implying its protective properties [27,28,47,51].

In contrast, Pihlstrøm et al. [25] observed no significant reduction/prevention of albuminuria, nor any influence on allograft gene expression. No significant group differences were found in the development of fibrosis, or chronic allograft damage index between the control and treated groups. Finally, only one study provided data regarding significant changes in the urinary peptidome of kidney transplant patients treated with paricalcitiol. However, these changes were not unidirectional, and some peptide levels were observed to increase, whereas others decreased [49]. According to the literature, the proven safety of paricalcitol legitimises its common use. Despite the possible occurrence of side effects, this drug is well-tolerated and it is considered to be safer than other alternatives [52,53].

3.3. Cinacalcet

The subtotal parathyroidectomy was the standard treatment of persistent hyperparathyroidism with hypercalcaemia in kidney transplant recipients. Currently, it has been replaced by the calcimimetic cinacalcet. In their publication, Soliman et al. aimed to prove the advantage of surgery over pharmacotherapy; however, no difference was found between the two groups in terms of the analysed parameters [54]. Furthermore, Mogl et al. compared the results of parathyroidectomy performed in patients receiving cinacalcet and those who were not treated with this drug. No difference was observed in the recurrence, or persistence of hyperparathyroidism, surgery duration, hospitalisation, or complications rate [55]. Calcimimetic agents represent a therapeutic alternative in transplant patients with persistent hyperparathyroidism and hypercalcaemia. They inhibit parathyroid hormone (PTH) secretion, increasing the sensitivity of the calcium-sensitive receptor in the parathyroid gland [56]. Some studies suggest this mechanism may be independent of PTH serum level suppression. The increase in bone-specific alkaline phosphatases, biochemical markers of bone accretion and a significant decrease in fasting urine calcium suggest the possibility of a beneficial impact of cinacalcet on bone remodelling [57]. The studies conducted by Pinho et al. demonstrated that all patients exhibited a significant reduction in parathyroid hormone levels iPTH from a mean value of 242.04 $\pm$105.82 pg/mL to a mean value of 145.62 $\pm$ 54.99 pg/mL. In addition, serum calcium levels normalized during the study period, from 11.16 mg/d to 9.95 mg/dL [29]. Vaquero et al., in their paper referring to the effect of cinacalcet on iPTH serum levels, showed that the drug administrated once daily reduced PTH in renal transplant recipients without holding PTH for 24 h. They noticed a progressive increase to a similar level to baseline 24 h after the administration [58]. Following a discontinuation of cinacalcet, calcaemia persisted at normal levels in 50%, although the drug had to be reintroduced in the other 50% after 10 $\pm$ 7.9 months. Nonetheless, there are currently no indicators suggesting in which group of patients further treatment may be discontinues [59]. As pointed out by Tillmann et al., the dysfunction of the parathyroid glands before the transplantation is associated with a clinically significant hyperparathyroidism after surgery. According to their study, PTH levels at 4 weeks post-transplantation may serve as a marker for the occurrence of hypercalcemic hyperparathyroidism during the follow-up [43]. Discontinuing cinacalcet within the first month following kidney transplantation frequently results in hypercalcaemia [60], although it is not a risk factor for allograft calcifications in the early post-transplant period. According to Paschoalin et al., allograft calcification was found in serial protocol biopsies after transplantation among patients receiving cinacalcet on dialysis, which was discontinued after the surgery. In fact, all biopsies showed nephrocalcinosis: either intratubular calcifications or calcifications in the parenchyma [22]. Furthermore, in patients with normocalcaemic persistent secondary hyperparathyroidism after kidney transplantation, cinacalcet improved the control of serum PTH without causing calcaemia, or phosphataemia. Observations were conducted on a group of 32 patients. Treatment with cinacalcet began after 16 months following KTx (median dose of 30 mg/day). Levels of iPTH decreased from a median of 364 pg/mL at the start of the study to 187 after 6 months (48.6% reduction) and to 145 after 12 months (60.2% reduction), without any changes in calcium and phosphorus [30]. Niel et al. in their study suggested that cinacalcet could be used in a paediatric population with an effect comparable to that observed in adults. The authors reported a case of a child after bilateral nephrectomy due to nephroblastoma. After renal transplantation, ionized serum calcium levels remained increased (1.6 mmol/L, 6.4 mg/dL), iPTH levels were high (348 ng/L), whereas serum phosphorus levels were low (1.24 mmol/L, 3.83 mg/dL). At that time, the patient received phosphate salts. A renal biopsy revealed traces of calcium deposits, and the ultrasonography showed a 9 mm $\times$ 4 mm parathyroid adenoma. Cinacalcet was introduced 4 months following renal transplantation; the initial dose was 15 mg/day, increased every 15 days up to 60 mg/day. At 7 months after renal transplantation, total serum calcium levels were normal (2.37 mmol/L, 9.48 mg/dL), as well as ionized serum calcium levels (1.24 mmol/L, 4.96 mg/dL) and serum phosphorus (1.42 mmol/L, 4.39 mg/dL); additionally, iPTH decreased (294 ng/L). A second renal biopsy performed as a control showed no traces of calcium deposits. Cervical ultrasonography performed 9 months after renal transplantation demonstrated no visible parathyroid hypertrophy, which was confirmed by another ultrasonography performed 6 months later [61]. In the study by Torregros et al. including 193 subjects (n = 193), a long-term effect of cinacalcet was verified. It was a retrospective study based on data from 17 renal transplant units from Spain. After cinacalcet pharmacotherapy mean calcium levels and iPTH decreased, whereas the mean phosphorus level was increased. The effects were maintained for up to 3 years. No changes were observed in the renal function or calcineurin inhibitor levels. Only 4.1% of patients discontinued cinacalcet due to intolerance, and 1.0% due to the lack of efficacy [31]. Zavvos et al. made similar observations based on a 5-years-long follow-up performed on a group of 47 patients. Adverse reactions were observed in four patients, comprising mild gastrointestinal complaints [32]. Satisfactory results of the therapy over 53 $\pm$ 7.4 months are also described by Paschoalin et al. [33]. There was no statistical difference in the percentage change in bone mineral density (BMD) at the femoral neck between the cinacalcet and placebo groups [34]. However, the issue emerged as to whether patients with hyperparathyroidism who received cinacalcet could be at increased risk of renal calcium deposits due to hypercalciuria and a subsequent renal transplant dysfunction. Seager et al. in their article described the first well-documented case in which cinacalcet contributed to the development of new renal calculi in a post-transplant patient with hyperparathyroidism (iPTH 346 pg/mL), hypercalcaemia (11.3 mg/dL), and good renal function (1.45 mg/dL). Interval imaging tracked the new onset of renal allograft stone formation after initiating cinacalcet up to 60 mg daily, which was accompanied by persistent hypercalciuria (478.2 mg/24 h). The nephrolithiasis resolved after discontinuing cinacalcet and a subtotal parathyroidectomy. This case emphasised the significance of interval monitoring of urinary calcium excretion and imaging of the transplanted kidney in terms of the recipients treated with cinacalcet for hyperparathyroidism after renal transplantation [62]. In turn, Seikrit et al. described a case of fulminant graft failure due to extensive tubular calcinosis after initiation of a calcimimetic therapy [63].

3.4. Parathyreidectomy

The timing of parathyroidectomy in kidney transplant candidates suffering from secondary hyperparathyroidism before transplantation, versus early or late after transplantation, remains controversial. Kovács et al. in their publication emphasised that if the patient was already waiting for a kidney transplant, it was worth performing the parathyroid surgery prior to it [64]. Conversely, Ivarsson et al. reported that parathyroidectomy was associated with improved survival in patients on maintenance dialysis, although not in patients with a renal allograft [65]. Li et al. in their study investigated the ultrasound-guided microwave ablation (MWA) in the treatment of patients with secondary hyperparathyroidism after a renal transplantation. They demonstrated that it was a safe and effective technique for destroying parathyroid gland tissue. Furthermore, the clinical effects of this intervention were long-lasting [66]. Nevertheless, a possible dangerous consequence of parathyroidectomy on kidney transplant function was also reported. In their publication, Ferreira et al. presented a comparison between acute and long-term renal changes following total parathyroidectomy with those occurring after general or urological surgery. They found that kidney filter function decreased in the acute period after parathyroidectomy, but then it returned to normal. Renal function was therefore similar to both preoperative function and to that of a control group of kidney-transplanted patients who had undergone other general surgery procedures [67]. Chudzinski et al. also reported that in the early postoperative period following parathyroidectomy, transient reductions in graft function occurred, although the procedure did not significantly impair the transplanted kidney function. eGFR was significantly lower on days 2 to 3 compared to the pre-parathyroidectomy values (52.38 vs. 44.89 mL/min/1.73 m^2; $p < 0.001$) [68]. In turn, Bures et al. observed no correlation between the postoperative iPTH and kidney function. They concluded that parathyroidectomy did not present negative effects on graft function, whether performed before or after (early or late) kidney transplantation [69]. Furthermore, Littbarski et al. in their study indicated that parathyroidectomy should be conducted before transplantation or after the first post-transplant year. Donor kidney function KDIGO stage III, blood group 0, and post-transplant parathyroidectomy were reported as independent significant risk factors for a compromised renal graft function in the short-term follow-up [70]. El-Husseini et al. aimed to evaluate the relationships of intraoperative iPTH with long-term iPTH levels post-parathyroidectomy in dialysis, as well as in renal transplant patients. According to their observations, intraoperative PTH after 20 min of the procedure is a good indicator of long-term PTH levels in dialysis and renal transplant patients [71]. In terms of the consequences of parathyroidectomy, it is vital to bear in mind that total parathyroidectomy increased the risk for acute hypocalcaemia after KTx, although it had no effect on long-term calcium homeostasis. Authors hypothesised that the early state of hypocalcaemia was due to the high-dose glucocorticoids required for induction, in addition to the preoperative undetectable PTH [72].

3.5. Etelcalcetide

A very interesting pharmacotherapeutic drug is etelcalcetide. Currently, there are no studies with this medicine in patients following kidney transplantation; however, there are results from studies in dialysis patients. Block GA et al. [73] conducted a randomized study evaluating the efficacy of etelcalcetide administered intravenously 3 times a week (n = 340), and patients were also treated with oral placebo daily. Cinacalcet was administered daily p.o. (n = 343) and placebo intravenously also 3 times a week during dialysis. The study was conducted for 26 weeks. PTH concentration reductions of more than 30% were achieved in the etelcalcetide group in 68.2% of patients vs. 57.7% for cinacalcetide. PTH concentration reductions of more than 50% were achieved in the etelcalcetide group in 52.4% of patients vs. 40.2% for cinacalcetide ($p = 0.001$). It should be noted that etelcalcetide lowered blood calcium levels more (68.9% vs. 59.8%). In another multicentre study (n = 2596), the authors also showed that etelcalcetide was effective in lowering PTH and blood calcium levels [74]. This fact highlights the important role of etelcalcetide in patients following kidney transplantation who often have tertiary hyperparathyroidism. Unfortunately, the limitation of the use of the described drug is the intravenous form of administration. No studies have investigated the use of etelcalcetid in patients following kidney transplantation but it may be useful in seriously hypercalcemic post-transplant HPT patients. Further studies are needed to evaluate clinical outcomes as well as long-term efficacy and safety.

4. Conclusions

In conclusion, paricalcitol, cinacalcet, and parathyroidectomy considerably decreased the iPTH level in the studied groups. Taking into consideration the analysis of the data, as well as complications occurring in the post-transplant HPT, particularly those pertaining to the cardiovascular system mainly associated with hypercalcemia, we tried to determine the best therapeutic strategy. In terms of all the analysed medications, paricalcitol most significantly increased the serum calcium concentration. In the case of active forms of vitamin D, there is a risk of developing treatment-related hypercalcemia; therefore, the use of these drugs in persistent hyperparathyroidism following kidney transplantation is questionable. Calcitriol is less effective in lowering PTH levels; yet, it causes a smaller increase in calcium levels than paricalcitol. Cinacalcet and parathyroidectomy lowered calcium levels in renal transplant patients with hypercalcaemia. In contrast, only parathyroidectomy produced no recurrence of hypercalcaemia. In the case of parathyroidectomy, very few studies with hard endpoints exist. Additionally, there is a risk of transient hypocalcaemia due to the "hungry-bone" syndrome, and persistent hypoparathyroidism with sustained hypocalcaemia/low bone turnover. More studies with regard to the safety of parathyroidectomy in renal transplant patients are necessary. Finally, cinacalcet seems to be a good candidate in the treatment of post-transplant HPT; however, it should be noted that it increases the risk of low bone turnover. Furthermore, it is vital to consider the individualisation of therapy.

In terms of the limitations of our paper, it is worth mentioning that that calcitriol was not included in Tables 1 and 2 due to the lack of sufficient data. Moreover, we have included papers about paricalcitol in Table 2, even though some of them were based on normocalcaemic and normophosphataemic patients with mild hyperparathyroidism. Such parameters do not constitute indications for urgent treatment; nevertheless, paricalcitol alters their values. Therefore, parameters were included in the table. A significant limitation of the study is the inability to perform a meta-analysis for the described issue due to the lack of sufficient data.

Author Contributions: Conceptualisation, M.M., K.L. and I.I.-P.; methodology, M.M.; software, M.M.; validation, M.M. and K.L.; formal analysis, K.L. and I.I.-P.; investigation, M.M., O.T., A.B. and M.P.; resources, M.M., O.T., A.B. and M.P.; data curation, M.M.; writing—original draft preparation, M.M., O.T., A.B. and M.P.; writing—review and editing, K.L. and I.I.-P.; visualisation, M.M.; supervision, K.L. and I.I.-P.; project administration, M.M.; funding acquisition, I.I.-P. All authors have read and agreed to the published version of the manuscript.

Funding: This research received no external funding.

Institutional Review Board Statement: Not applicable.

Informed Consent Statement: Not applicable.

Data Availability Statement: Not applicable.

Conflicts of Interest: The authors declare no conflict of interest. The funders had no role in the design of the study; in the collection, analyses, or interpretation of data; in the writing of the manuscript; or in the decision to publish the results.

References

1. Silver, J.; Levi, R. Regulation of PTH synthesis and secretion relevant to the management of secondary hyperparathyroidism in chronic kidney disease. *Kidney Int. Suppl.* **2005**, *95*, S8–S12. [CrossRef] [PubMed]
2. van der Plas, W.Y.; Noltes, M.E.; van Ginhoven, T.M.; Kruijff, S. Secondary and Tertiary Hyperparathyroidism: A Narrative Review. *Scand. J. Surg.* **2020**, *109*, 271–278. [CrossRef] [PubMed]

3. Hruska, K.A.; Mathew, S.; Lund, R.; Qiu, P.; Pratt, R. Hyperphosphatemia of chronic kidney disease. *Kidney Int.* **2008**, *74*, 148–157. [CrossRef] [PubMed]
4. Yuen, N.K.; Ananthakrishnan, S.; Campbell, M.J. Hyperparathyroidism of Renal Disease. *Perm J.* **2016**, *20*, 15–127. [CrossRef] [PubMed]
5. Miedziaszczyk, M.; Idasiak-Piechocka, I.; Wiśniewski, O.W.; Lacka, K. A systematic review of the pharmacotherapy of secondary hyperparathyroidism (SHPT) in grades 3-5 Chronic Kidney Disease (CKD). *Eur Rev. Med. Pharmacol Sci.* **2022**, *26*, 232–239.
6. Delos Santos, R.; Rossi, A.; Coyne, D.; Maw, T.T. Management of Post-transplant Hyperparathyroidism and Bone Disease. *Drugs* **2019**, *79*, 501–513. [CrossRef]
7. Evenepoel, P.; Claes, K.; Kuypers, D.; Maes, B.; Bammens, B.; Vanrenterghem, Y. Natural history of parathyroid function and calcium metabolism after kidney transplantation: A single-centre study. *Nephrol Dial. Transplant.* **2004**, *19*, 1281–1287. [CrossRef]
8. Messa, P.; Sindici, C.; Cannella, G.; Miotti, V.; Risaliti, A.; Gropuzzo, M.; Di Loreto, P.L.; Bresadola, F.; Mioni, G. Persistent secondary hyperparathyroidism after renal transplantation. *Kidney Int.* **1998**, *54*, 1704–1713. [CrossRef]
9. Jeon, H.J.; Kim, Y.J.; Kwon, H.Y.; Koo, T.Y.; Baek, S.H.; Kim, H.J.; Huh, W.S.; Huh, K.H.; Kim, M.S.; Kim, Y.S.; et al. Impact of parathyroidectomy on allograft outcomes in kidney transplantation. *Transpl. Int. Off. J. Eur. Soc. Organ Transplant.* **2012**, *25*, 1248–1256. [CrossRef]
10. Nakamura, M.; Tanaka, K.; Marui, Y.; Tomikawa, S. Clinicopathological analysis of persistent hypercalcemia and hyperparathyroidism after kidney transplantation in long-term dialysis patients. *Ther Apher Dial.* **2013**, *17*, 551–556. [CrossRef]
11. Taniguchi, M.; Tokumoto, M.; Matsuo, D.; Motoyama, K.; Sugitani, A.; Kuroki, S.; Yotsueda, H.; Tsuruya, K.; Hirakata, H.; Iida, M. Persistent hyperparathyroidism in renal allograft recipients: Vitamin D receptor, calcium-sensing receptor, and apoptosis. *Kidney Int.* **2006**, *70*, 363–370. [CrossRef]
12. Garcia-Montemayor, V.; Sánchez-Agesta, M.; Agüera, M.L.; Calle, Ó.; Navarro, M.D.; Rodríguez, A.; Aljama, P. Influence of Pre-renal Transplant Secondary Hyperparathyroidism on Later Evolution After Transplantation. *Transplant. Proc.* **2019**, *51*, 344–349. [CrossRef] [PubMed]
13. Lou, I.; Foley, D.; Odorico, S.K.; Leverson, G.; Schneider, D.F.; Sippel, R.; Chen, H. How Well Does Renal Transplantation Cure Hyperparathyroidism? *Ann. Surg.* **2015**, *262*, 653–659. [CrossRef] [PubMed]
14. Yamamoto, T.; Tominaga, Y.; Okada, M.; Hiramitsu, T.; Tsujita, M.; Goto, N.; Narumi, S.; Watarai, Y. Characteristics of Persistent Hyperparathyroidism After Renal Transplantation. *World J. Surg.* **2016**, *40*, 600–606. [CrossRef] [PubMed]
15. Kirnap, N.G.; Kirnap, M.; Sayin, B.; Akdur, A.; Bascil Tutuncu, N.; Haberal, M. Risk Factors and Treatment Options for Persistent Hyperparathyroidism After Kidney Transplantation. *Transplant. Proc.* **2020**, *52*, 157–161. [CrossRef]
16. Triponez, F.; Clark, O.H.; Vanrenthergem, Y.; Evenepoel, P. Surgical treatment of persistent hyperparathyroidism after renal transplantation. *Ann. Surg.* **2008**, *248*, 18–30. [CrossRef]
17. Evenepoel, P. Recovery versus persistence of disordered mineral metabolism in kidney transplant recipients. *Semin Nephrol.* **2013**, *33*, 191–203. [CrossRef]
18. Gwinner, W.; Suppa, S.; Mengel, M.; Hoy, L.; Kreipe, H.H.; Haller, H.; Schwarz, A. Early calcification of renal allografts detected by protocol biopsies: Causes and clinical implications. *Am. J. Transplant.* **2005**, *5*, 1934–1941. [CrossRef]
19. Mazzaferro, S.; Pasquali, M.; Taggi, F.; Baldinelli, M.; Conte, C.; Muci, M.L.; Pirozzi, N.; Carbone, I.; Francone, M.; Pugliese, F. Progression of coronary artery calcification in renal transplantation and the role of secondary hyperparathyroidism and inflammation. *Clin. J. Am. Soc. Nephrol.* **2009**, *4*, 685–690. [CrossRef]
20. Rojas, E.; Carlini, R.G.; Clesca, P.; Arminio, A.; Suniaga, O.; De Elguezabal, K.; Weisinger, J.R.; Hruska, K.A.; Bellorin-Font, E. The pathogenesis of osteodystrophy after renal transplantation as detected by early alterations in bone remodeling. *Kidney Int.* **2003**, *63*, 1915–1923. [CrossRef]
21. Ivarsson, K.M.; Clyne, N.; Almquist, M.; Akaberi, S. Hyperparathyroidism and new onset diabetes after renal transplantation. *Transplant. Proc.* **2014**, *46*, 145–150. [CrossRef] [PubMed]
22. Paschoalin, R.P.; Torregrosa, J.V.; Barros, X.; Durán, C.E.; Soler, M.; Campistol, J.M. Withdrawal of cinacalcet at the time of renal transplantation is not a risk factor for allograft calcifications in the early posttransplantation period. *Transplant. Proc.* **2012**, *44*, 2379–2380. [CrossRef]
23. Julian, B.A.; Laskow, D.A.; Dubovsky, J.; Dubovsky, E.V.; Curtis, J.J.; Quarles, L.D. Rapid loss of vertebral mineral density after renal transplantation. *N. Eng. J. Med.* **1991**, *325*, 544–550. [CrossRef] [PubMed]
24. Hernández, D.; Rufino, M.; Bartolomei, S.; González-Rinne, A.; Lorenzo, V.; Cobo, M.; Torres, A. Clinical impact of preexisting vascular calcifications on mortality after renal transplantation. *Kidney Int.* **2005**, *67*, 2015–2020. [CrossRef] [PubMed]
25. Pihlstrøm, H.K.; Gatti, F.; Hammarström, C.; Eide, I.A.; Kasprzycka, M.; Wang, J.; Haraldsen, G.; Svensson, M.H.S.; Midtvedt, K.; Mjøen, G.; et al. Early introduction of oral paricalcitol in renal transplant recipients. An open-label randomized study. *Transpl Int.* **2017**, *30*, 827–840. [CrossRef] [PubMed]
26. Borrego Utiel, F.J.; Bravo Soto, J.A.; Merino Pérez, M.J.; González Carmelo, I.; López Jiménez, V.; García Álvarez, T.; Acosta Martínez, Y.; Mazuecos Blanca, M.A. Effect of paricalcitol on mineral bone metabolism in kidney transplant recipients with secondary hyperparathyroidism. *Nefrologia* **2015**, *35*, 363–373, (In English & Spanish). [CrossRef] [PubMed]
27. Trillini, M.; Cortinovis, M.; Ruggenenti, P.; Reyes Loaeza, J.; Courville, K.; Ferrer-Siles, C.; Prandini, S.; Gaspari, F.; Cannata, A.; Villa, A.; et al. Paricalcitol for secondary hyperparathyroidism in renal transplantation. *J. Am. Soc. Nephrol.* **2015**, *26*, 1205–1214. [CrossRef] [PubMed]

28. Gonzalez, E.; Rojas-Rivera, J.; Polanco, N.; Morales, E.; Morales, J.M.; Egido, J.; Amado, A.; Praga, M. Effects of oral paricalcitol on secondary hyperparathyroidism and proteinuria of kidney transplant patients. *Transplantation* **2013**, *95*, e49–e52. [CrossRef]
29. Pinho, L.R.; Ribeiro Santos, M.J.; Pestana Vasconcelos, M. Cinacalcet in the treatment of persistent hyperparathyroidism after kidney transplantation. *Clin. Nephrol.* **2011**, *75*, 263–268. [CrossRef]
30. Torregrosa, J.V.; Morales, E.; Díaz, J.M.; Crespo, J.; Bravo, J.; Gómez, G.; Gentil, M.A.; Rodríguez-Benot, A.; Rodríguez García, M.; López-Jiménez, V.; et al. Cinacalcet en el manejo del hiperparatiroidismo secundario normocalcémico tras el trasplante renal: Estudio multicéntrico de un año de seguimiento [Cinacalcet in the management of normocalcaemic secondary hyperparathyroidism after kidney transplantation: One-year follow-up multicentre study]. *Nefrologia* **2014**, *34*, 62–68. (In Spanish
31. Torregrosa, J.V.; Morales, E.; Díaz, J.M.; Crespo, J.; Bravo, J.; Gómez, G.; Gentil, M.Á.; Rodríguez Benot, A.; García, M.R.; Jiménez, V.L.; et al. Cinacalcet for hypercalcaemic secondary hyperparathyroidism after renal transplantation: A multicentre, retrospective, 3-year study. *Nephrology* **2014**, *19*, 84–93. [CrossRef] [PubMed]
32. Zavvos, V.; Fyssa, L.; Papasotiriou, M.; Papachristou, E.; Ntrinias, T.; Savvidaki, E.; Goumenos, D.S. Long-Term Use of Cinacalcet in Kidney Transplant Recipients With Hypercalcemic Secondary Hyperparathyroidism: A Single-Center Prospective Study. *Exp. Clin. Transplant.* **2018**, *16*, 287–293. [PubMed]
33. Paschoalin, R.P.; Torregrosa, J.V.; Sánchez-Escuredo, A.; Barros, X.; Durán, C.E.; Campistol, J.M. Cinacalcet treatment for stable kidney transplantation patients with hypercalcemia due to persistent secondary hyperparathyroidism: A long-term follow-up. *Transplant. Proc.* **2012**, *44*, 2588–2589. [CrossRef] [PubMed]
34. Evenepoel, P.; Cooper, K.; Holdaas, H.; Messa, P.; Mourad, G.; Olgaard, K.; Rutkowski, B.; Schaefer, H.; Deng, H.; Torregrosa, J.V.; et al. A randomized study evaluating cinacalcet to treat hypercalcemia in renal transplant recipients with persistent hyperparathyroidism. *Am. J. Transplant.* **2014**, *14*, 2545–2555. [CrossRef] [PubMed]
35. Donate-Correa, J.; Henríquez-Palop, F.; Martín-Núñez, E.; Hernández-Carballo, C.; Ferri, C.; Pérez-Delgado, N.; Muros-de Fuentes, M.; Mora-Fernández, C.; Navarro-González, J.F. Anti-inflammatory profile of paricalcitol in kidney transplant recipient *Nefrologia* **2017**, *37*, 622–629, (In English & Spanish). [CrossRef] [PubMed]
36. Pérez, V.; Sánchez-Escuredo, A.; Lauzurica, R.; Bayés, B.; Navarro-Muñoz, M.; Pastor, M.C.; Cañas, L.; Bonet, J.; Romero, R. Magnetic bead-based proteomic technology to study paricalcitol effect in kidney transplant recipients. *Eur. J. Pharmacol.* **201** *709*, 72–79. [CrossRef]
37. Moreno, P.; Coloma, A.; Torregrosa, J.V.; Montero, N.; Francos, J.; Codina, S.; Manonelles, A.; Bestard, O.; García-Barrasa, A Melilli, E.; et al. Long-term results of a randomized study comparing parathyroidectomy with cinacalcet for treating tertiar hyperparathyroidism. *Clin. Transplant.* **2020**, *34*, e13988. [CrossRef]
38. Cruzado, J.M.; Moreno, P.; Torregrosa, J.V.; Taco, O.; Mast, R.; Gómez-Vaquero, C.; Polo, C.; Revuelta, I.; Francos, J.; Torras, J.; et a A Randomized Study Comparing Parathyroidectomy with Cinacalcet for Treating Hypercalcemia in Kidney Allograft Recipients with Hyperparathyroidism. *J. Am. Soc. Nephrol.* **2016**, *27*, 2487–2494. [CrossRef]
39. Cianciolo, G.; Galassi, A.; Capelli, I.; Angelini, M.L.; La Manna, G.; Cozzolino, M. Vitamin D in Kidney Transplant Recipient Mechanisms and Therapy. *Am. J. Nephrol.* **2016**, *43*, 397–407. [CrossRef]
40. Rodriguez, M.; Munoz-Castaneda, J.R.; Almaden, Y. Therapeutic use of calcitriol. *Curr Vasc Pharmacol.* **2014**, *12*, 294–29 [CrossRef]
41. Torregrosa, J.V.; Fuster, D.; Duran, C.E.; Oppenheimer, F.; Muxí, Á.; Rubello, D.; Pons, F.; Campistol, J.M. Set point of calcium i severe secondary hyperparathyroidism is altered and does not change after successful kidney transplantation. *Endocrine* **2015**, *4* 709–711. [CrossRef] [PubMed]
42. Torregrosa, J.V.; Barros, X. Management of hypercalcemia after renal transplantation. *Nefrologia* **2013**, *33*, 751–757, (In English Spanish). [PubMed]
43. Tillmann, F.P.; Wächtler, C.; Hansen, A.; Rump, L.C.; Quack, I. Vitamin D and cinacalcet administration pre-transplantation predict hypercalcaemic hyperparathyroidism post-transplantation: A case-control study of 355 deceased-donor renal transplant recipients over 3 years. *Transplant. Res.* **2014**, *3*, 21. [CrossRef] [PubMed]
44. Haffner, D.; Leifheit-Nestler, M. CKD-MBD post kidney transplantation. *Pediatr Nephrol.* **2021**, *36*, 41–50. [CrossRef]
45. Wada, Y.; Iyoda, M.; Iseri, K.; Arai-Nunota, N.; Saito, T.; Hamada, T.; Tachibana, S.; Ikeda, M.; Shibata, T. Combination Therapy of Denosumab and Calcitriol for a Renal Transplant Recipient with Severe Bone Loss due to Therapy-Resistant Hyperparathyroidism *Tohoku J. Exp. Med.* **2016**, *238*, 205–212. [CrossRef]
46. Robinson, D.M.; Scott, L.J. Paricalcitol: A review of its use in the management of secondary hyperparathyroidism. *Drugs* **2005**, *6* 559–576. [CrossRef]
47. Žilinská, Z.; Dedinská, I.; Breza, J.; Laca, L. Effect of Paricalcitol on Bone Density After Kidney Transplantation: Analysis of Transplant Centers. *Iran. J. Kidney Dis.* **2017**, *11*, 461–466.
48. Donate-Correa, J.; Henríquez-Palop, F.; Martín-Núñez, E.; Pérez-Delgado, N.; Muros-de-Fuentes, M.; Mora-Fernández, C Navarro-González, J.F. Effect of Paricalcitol on FGF-23 and Klotho in Kidney Transplant Recipients. *Transplantation* **2016**, *1C* 2432–2438. [CrossRef]
49. Pérez, V.; Sánchez, A.; Bayés, B.; Navarro-Muñoz, M.; Lauzurica, R.; Pastor, M.C.; Romero, R. Effect of paricalcitol on the urinary peptidome of kidney transplant patients. *Transplant. Proc.* **2010**, *42*, 2924–2927. [CrossRef] [PubMed]
50. Hu, M.C.; Kuro-o, M.; Moe, O.W. Klotho and chronic kidney disease. *Contrib. Nephrol.* **2013**, *180*, 47–63.

51. Amer, H.; Griffin, M.D.; Stegall, M.D.; Cosio, F.G.; Park, W.D.; Kremers, W.K.; Heilman, R.L.; Mazur, M.J.; Hamawi, K.; Larson, T.S.; et al. Oral paricalcitol reduces the prevalence of posttransplant hyperparathyroidism: Results of an open label randomized trial. *Am. J. Transplant.* **2013**, *13*, 1576–1585. [CrossRef] [PubMed]
52. Cheng, J.; Zhang, W.; Zhang, X.; Li, X.; Chen, J. Efficacy and safety of paricalcitol therapy for chronic kidney disease: A meta-analysis. *Clin. J. Am. Soc. Nephrol.* **2012**, *7*, 391–400. [CrossRef] [PubMed]
53. Dyer, C.A. Safety and tolerability of paricalcitol in patients with chronic kidney disease. *Expert Opin. Drug Saf.* **2013**, *12*, 717–728. [CrossRef] [PubMed]
54. Soliman, A.R.; Maamoun, H.A.; Soliman, M.A.; Darwish, H.; Elbanna, E. Cinacalcet versus Parathyroidectomy in the Treatment of Secondary Hyperparathyroidism Post Renal Transplantation. *Rom. J. Intern. Med.* **2016**, *54*, 184–189. [CrossRef]
55. Mogl, M.T.; Skachko, T.; Dobrindt, E.M.; Reinke, P.; Bures, C.; Pratschke, J.; Rayes, N. Surgery for Renal Hyperparathyroidism in the Era of Cinacalcet: A Single-Center Experience. *Scand. J. Surg.* **2021**, *110*, 66–72. [CrossRef]
56. Guerra, R.; Auyanet, I.; Fernández, E.J.; Pérez, M.Á.; Bosch, E.; Ramírez, A.; Suria, S.; Checa, M.D. Hypercalcemia secondary to persistent hyperparathyroidism in kidney transplant patients: Analysis after a year with cinacalcet. *J. Nephrol.* **2011**, *24*, 78–82. [CrossRef]
57. Boulanger, H.; Haymann, J.P.; Fouqueray, B.; Mansouri, R.; Metivier, F.; Mercadal, L.; Attaf, D.; Flamant, M.; Glotz, D. Effet du cinacalcet sur l'homéostasie calcique et le remodelage osseux chez 13 transplantés rénaux présentant une hyperparathyroïdie avec hypercalcémie [Cinacalcet impact on calcium homeostasis and bone remodeling in 13 renal transplanted patients with hyperparathyroidism and hypercalcaemia]. *Nephrol Ther.* **2012**, *8*, 47–53. (In French)
58. Vaquero, E.; Esteban de la Rosa, R.J.; Oliva, N.; Fernández Castillo, R.; Fernández Gallegos, R.; Bravo Soto, J. Efecto de cinacalcet sobre la hormona paratiroidea en el hiperparatiroidismo hipercalcémico del trasplantado renal [Effect of cinacalcet on parathyroid hormone level in hypercalcemic hyperparathyroidism of patients with renal transplantation]. *Med. Clin.* **2011**, *138*, 323–326. [CrossRef]
59. Paschoalin, R.P.; Torregrosa, J.V.; Barros, X.; Durán, C.E.; Campistol, J.M. Cinacalcet de novo in persistent hypercalcemia after kidney transplantation secondary to hyperparathyroidism: Long-term follow-up and effect of withdrawal. *Transplant. Proc.* **2012**, *44*, 2376–2378. [CrossRef]
60. Mawad, H.; Bouchard, H.; Tran, D.; Ouimet, D.; Lafrance, J.P.; Bell, R.Z.; Bezzaoucha, S.; Boucher, A.; Collette, S.; Pichette, V.; et al. Retrospective Study Looking at Cinacalcet in the Management of Hyperparathyroidism after Kidney Transplantation. *J. Transplant.* **2017**, *2017*, 8720283. [CrossRef]
61. Niel, O.; Maisin, A.; Macher, M.A.; Peuchmaur, M.; Deschênes, G. Cinacalcet in hyperparathyroidism management after pediatric renal transplantation. *CEN Case Rep.* **2016**, *5*, 141–143. [CrossRef] [PubMed]
62. Seager, C.M.; Srinivas, T.R.; Flechner, S.M. Development of nephrolithiasis in a renal transplant patient during treatment with Cinacalcet. *Ann. Transplant.* **2013**, *18*, 31–35. [CrossRef]
63. Seikrit, C.; Mühlfeld, A.; Groene, H.J.; Floege, J. Renal allograft failure in a hyperparathyroid patient following initiation of a calcimimetic. *Nat. Rev. Nephrol.* **2011**, *7*, 237–241. [CrossRef] [PubMed]
64. Kovács, D.Á.; Fedor, R.; Asztalos, L.; Andrási, M.; Szabó, R.P.; Kanyári, Z.; Barna, S.; Nemes, B.; Győry, F. Surgical Treatment of Hyperparathyroidism After Kidney Transplant. *Transplant. Proc.* **2019**, *51*, 1244–1247. [CrossRef] [PubMed]
65. Ivarsson, K.M.; Akaberi, S.; Isaksson, E.; Reihnér, E.; Rylance, R.; Prütz, K.G.; Clyne, N.; Almquist, M. The effect of parathyroidectomy on patient survival in secondary hyperparathyroidism. *Nephrol Dial. Transplant.* **2015**, *30*, 2027–2033. [CrossRef]
66. Li, X.; An, C.; Yu, M.; Peng, L. US-guided microwave ablation for secondary hyperparathyroidism in patients after renal transplantation: A pilot study. *Int. J. Hyperthermia.* **2019**, *36*, 322–327. [CrossRef]
67. Ferreira, G.F.; Montenegro, F.L.; Machado, D.J.; Ianhez, L.E.; Nahas, W.C.; David-Neto, E. Parathyroidectomy after kidney transplantation: Short-and long-term impact on renal function. *Clinics* **2011**, *66*, 431–435. [CrossRef]
68. Chudzinski, W.; Wyrzykowska, M.; Nazarewski, S.; Durlik, M.; Galazka, Z. Does the Parathyroidectomy Endanger the Transplanted Kidney? *Transplant. Proc.* **2016**, *48*, 1633–1636. [CrossRef]
69. Bures, C.; Seika, P.; Skachko, T.; Dobrindt, E.M.; Rayes, N.; Pratschke, J.; Goretzki, P.E.; Mogl, M.T. Influence of Parathyroidectomy on Kidney Graft Function in Secondary and Tertiary Hyperparathyroidism. *Transplant. Proc.* **2020**, *52*, 3134–3143. [CrossRef]
70. Littbarski, S.A.; Kaltenborn, A.; Gwiasda, J.; Beneke, J.; Arelin, V.; Schwager, Y.; Stupak, J.V.; Marcheel, I.L.; Emmanouilidis, N.; Jäger, M.D.; et al. Timing of parathyroidectomy in kidney transplant candidates with secondary hyperparathryroidism: Effect of pretransplant versus early or late post-transplant parathyroidectomy. *Surgery* **2018**, *163*, 373–380. [CrossRef]
71. El-Husseini, A.; Wang, K.; Edon, A.; Saxon, D.; Lima, F.; Sloan, D.; Sawaya, B.P. Value of Intraoperative Parathyroid Hormone Assay during Parathyroidectomy in Dialysis and Renal Transplant Patients with Secondary and Tertiary Hyperparathyroidism. *Nephron* **2018**, *138*, 119–128. [CrossRef] [PubMed]
72. Simons, M.; Bautista, J.; Occhiogrosso, R.; Scott-Sheldon, L.A.; Gohh, R. Acute hypocalcemia following kidney transplantation may depend on the type of remote parathyroidectomy: A retrospective cohort study. *Clin. Nephrol.* **2017**, *87*, 287–292. [CrossRef] [PubMed]
73. Block, G.A.; Bushinsky, D.A.; Cheng, S.; Cunningham, J.; Dehmel, B.; Drueke, T.B.; Ketteler, M.; Kewalramani, R.; Martin, K.J.; Moe, S.M.; et al. Effect of Etelcalcetide vs Cinacalcet on Serum Parathyroid Hormone in Patients Receiving Hemodialysis With Secondary Hyperparathyroidism: A Randomized Clinical Trial. *JAMA* **2017**, *317*, 156–164. [CrossRef] [PubMed]

74. Karaboyas, A.; Muenz, D.; Fuller, D.S.; Desai, P.; Lin, T.C.; Robinson, B.M.; Rossetti, S.; Pisoni, R.L. Etelcalcetide Utilization, Dosing Titration, and Chronic Kidney Disease-Mineral and Bone Disease (CKD-MBD) Marker Responses in US Hemodialysis Patients. *Am. J. Kidney Dis.* **2022**, *79*, 362–373. [CrossRef]

biomedicines

MDPI

Systematic Review

Chronic Hepatitis B and Related Liver Diseases Are Associated with Reduced 25-Hydroxy-Vitamin D Levels: A Systematic Review and Meta-Analysis

Anindita Banerjee [1,*], Shreyasi Athalye [1], Naveen Khargekar [2], Poonam Shingade [3] and Manisha Madkaikar [4]

1 Transfusion Transmitted Disease Department, ICMR-National Institute of Immunohaematology, Parel, Mumbai 400012, India
2 Hematogenetics Department, ICMR-National Institute of Immunohaematology, Parel, Mumbai 400012, India
3 Department of Community Medicine, ESIC Medical College, Gulbarga 585106, India
4 Pediatric Immunology & Leukocyte Biology Department, ICMR-National Institute of Immunohaematology, Parel, Mumbai 400012, India
* Correspondence: anny.banerjee@gmail.com; Tel.: +91-22-2413-8518/19

Abstract: Hepatitis B infection is a major public health problem globally leading to chronic liver disease and death, which are influenced by various environmental and host factors including serum 25-hydroxy-vitamin D levels. There is no comprehensive systematic review reporting the association of serum 25-hydroxy-vitamin D levels and different stages of chronic hepatitis B. This study aimed to analyze the association of 25-hydroxy-vitamin D levels in chronic hepatitis B with various determinants and outcomes. A bibliographic search in PubMed, Google Scholar, and Scopus was conducted using the search terms "Vitamin D", "cholecalciferol", "calcitriol", "Hepatitis B", and "HBV", which were published until September 2022. Meta-analysis using the "metafor" package in R was conducted with a random effect model. This analysis included 33 studies with 6360 chronic hepatitis B patients. The pooled estimates of serum 25-hydroxy-vitamin D level among CHB cases was 21.05 ng/mL and was significantly lower compared to healthy controls. ($p < 0.005$). Reduced serum 25-hydroxy-vitamin D level was significantly associated with the severity of liver fibrosis as well as HBe positivity. This analysis suggests that serum 25-hydroxy-vitamin D levels are associated with disease activity and pathobiology, although the exact nature of the cause–effect relationship cannot be discerned from this study.

Keywords: vitamin D; chronic hepatitis B; chronic liver disease; 25-hydroxy-vitamin D; fibrosis; cirrhosis; meta-analysis

1. Introduction

Hepatitis B infection is a public health problem affecting nearly one third of the population worldwide. In some of the infected individuals, hepatitis B infection persists as chronic infection, which later leads to complications such as hepatic fibrosis, decompensated cirrhosis, and hepatocellular carcinoma [1]. The progression of hepatitis B infection to chronic infection and liver decomposition is regulated by various host and environmental factors [1,2]. The host immune factors are controlled by nutrition, endocrine, and other determinants [1].Vitamin D is one such molecule with multiple effects on immunity, inflammation, and fibrosis. Vitamin D, a fat-soluble vitamin is crucial for a plethora of biological and physiological functions in the body. The biologically active form of vitamin D is Calcitriol (1, 25-dihydroxy vitamin D); however, the serum level of vitamin D is determined by the level of 25-hydroxy-vitamin D, which is an indicator of vitamin D sufficiency. While various classifications of vitamin D deficiency have been proposed, the widely accepted classification of vitamin D deficiency is as follows: Normal/optimum: >30 ng/mL; Insufficiency: >20 but <30 ng/mL; mild deficiency: >10 but <20 ng/mL; severe deficiency: <10 ng/mL [3]. Vitamin D3 or cholecalciferol is a hormone synthesized from the skin upon sun exposure and gets hydroxylated in the liver by 25-hydroxylase to be converted to 25-hydroxy-vitamin D followed by one more round of hydroxylation in the kidney to produce 1,25-dihydroxy-vitamin D or calcitriol, which is the physiologically active form of vitamin D. Calcitriol has a very short half-life and is not representative of the body's reserve of vitamin D levels, which is why serum 25-hydroxy-vitamin D levels are considered as a measure of the sufficiency of vitamin D. Epidemiological evidence have highlighted the role of vitamin D in the pathogenesis of several infectious, autoimmune, and malignant diseases. Vitamin D supplementation in a few of these disease cohorts has been found to be beneficial [4]. Experimental data shows that apart from bone and muscle health and calcium metabolism, vitamin D exerts direct anti-inflammatory, anti-viral and immunomodulatory effects through various cross-talks with the cellular targets, which are activated by vitamin D-VDR signaling [5,6]. It is a well-known fact that host-viral interaction and host-immune response are responsible for various HBV outcomes. Dysregulated immune response can lead to chronic infection and late complications. Interactive crosstalk between innate and adaptive immune response largely determines the ultimate control of HBV infection. Available immunological reports found that Toll-Like Receptors (TLRs), monocytes, natural killer T-cells, cytotoxic T lymphocyte (CTL), Th1 CD4+ T cells, and dendritic cells play an essential role in the fate of HBV infection [7,8]. As vitamin D has an immunomodulatory role, it is presumable that host vitamin D levels might determine the outcome of HBV infection. Lower levels of vitamin D have been reported to be associated with chronic liver disease [9]. Moreover, one study from Israel found that patients with chronic hepatitis C with higher vitamin D levels demonstrated better virological response compared to those with lower levels [10]. Previous studies on chronic hepatitis B patients also found patients with HBV infections had significantly low levels of vitamin D and HBV DNA levels correlated inversely with the serum vitamin D levels [11]. Similarly, one meta-analysis of seven such studies found that chronic hepatitis B patients had low vitamin D levels and it correlated with the viral load [12]. However, the number of included studies was very small, with most studies from Asia. Furthermore, the study did not explore the association of vitamin D levels with disease status or other variables, which can strongly influence the association. Thus, it remains to be determined whether serum 25-hydroxy-vitamin D levels differ significantly among chronic hepatitis B patients and influence the various outcomes. Therefore, the aim of the study is to provide a detailed account of the role of 25-hydroxy-vitamin D levels in the context of chronic hepatitis B with various determinants and outcomes.

2. Materials and Methods

2.1. Protocol and Registration

The Preferred Reporting Items for Systematic Reviews and Meta-Analysis (PRISMA) 2020 guidelines were followed throughout this systematic review and meta-analysis [13]. The review protocol was registered in the International prospective register of systematic reviews, PROSPERO. (https://www.crd.york.ac.uk/prospero/display_record.php?RecordID=366783, Registration number: CRD42022366783, accessed on 15 December 2022).

2.2. Search Strategy

We conducted a search for studies published from January 2000 to September 2022 in PubMed, Google Scholar, and Scopus. Search terms used for identifying publications included "Vitamin D", "Hepatitis B", "Calcitriol", "Cholecalciferol", and "HBV". The database search strategy and results are provided in Table 1.

Table 1. Database search strategy and results.

Sr. No.	Key Words	Databases		
		PubMed	**Google Scholar**	**Scopus**
1	"Calcitriol" and "Hepatitis B"	44	998	132
2	"Cholecalciferol" and "Hepatitis B"	6	999	8
3	"Vitamin D" and "Hepatitis B"	192	998	200
4	"Calcitriol" and "HBV"	26	996	32
5	"Cholecalciferol" and "HBV"	2	606	3
6	"Vitamin D" and "HBV"	91	995	130

The publications were uploaded in Rayyan software (https://rayyan.ai/, accessed on 15 December 2022) to remove the duplicates and screen the articles according to the inclusion criteria [14].

2.3. Study Inclusion/Exclusion Criteria as per PICOS

2.3.1. Inclusion Criteria

The following inclusion criteria were adopted to screen the articles:

1. Individuals with hepatitis B infection (inactive carriers, chronic hepatitis);
2. Adult participants >18 years of age;
3. Classified the serum vitamin D concentration as either mean Vitamin D (nmol/L; ng/mL);
4. Articles reporting hepatitis B infection with cirrhosis/fibrosis/HCC;
5. In the English language;
6. Full text available;
7. Retrospective studies, cross-sectional, cohort, or randomized controlled trials.

2.3.2. Exclusion Criteria

1. We excluded articles that were in a foreign language, or which did not contain the relevant information as per the inclusion criteria;
2. Studies including pregnant women and children were not included;
3. We also excluded literature reviews, editorial reviews, and systematic reviews;
4. Editorials, brief communications, and conference proceedings were also excluded;
5. Articles estimating the values of only 1, 25-dihydroxy vitamin D were also excluded.

2.4. Evaluation of the Methodological Quality of the Studies Included

The methodological quality was independently assessed by two reviewers (SA and NK) using Rayyan Software [14]. Discussion with a third reviewer (AB) was performed to resolve the disagreements and conflicts. The quality assessment was performed using the Downs and Black checklist for quality assessment [15]. The tool considers study-related characteristics of quality, external validity, biases such as study and selection bias, confounding variable consideration and power to assess the overall quality. One point (yes) or zero (no) was scored for each item, excluding the power question. The power was scored on a 6-point scale, and the post-hoc power calculations were performed using G*power software [16]. The total score determined an overall quality index, which was used to classify the studies as excellent (>25), good (18–25), fair (13–17) and poor (<13).

2.5. Data Extraction

SA and NK extracted the data using a standardized data format from studies that gave the number of cases according to 25-hydroxy-vitamin D levels and hepatitis B infection/outcomes. Research articles with inadequate or unclear results were excluded from quantitative analysis. Any discrepancies were resolved by through detailed discussion and consensus between the two authors (SA and NK) and an independent review by the third reviewer (AB). An electronic spreadsheet was created in which the following information was recorded: authors, year of publication, country where the study was conducted, type of publication, study design, sample size, gender, age, and 25-hydroxy-vitamin D levels (in cases and controls, in case of case-control studies).

2.6. Statistical Analysis

Data analysis was performed using the 'meta' and 'metafor' package in RStudio 2022.07.2 Build 576 with R for Windows version 4.2.1. The packages contain functions to estimate the common effect and random effects, generate meta-analytical plots such as forest plots, funnel plots, as well as sub-group and meta-regression analysis. A mean difference with a 95% confidence interval was used to determine the difference between the serum 25-hydroxy-vitamin D levels between chronic hepatitis B patients and controls. Cochran Q test and I^2 statistics were used to estimate the heterogeneity between the studies. The percentage of variation across studies leading to heterogeneity rather than chance was defined as low, moderate, or high for values of 25%, 50%, and 75%, respectively. The publication bias was assessed using Funnel plots and Egger's test.

3. Results

3.1. Search Results and Study Selection

A total of 6458 articles were identified through the search strategy, out of which 3297 were removed as duplicates and 3120 were excluded after screening titles and abstracts. Full-text review of 40 articles was performed, out of which 7 articles were excluded for not matching the eligibility criteria. A total of 33 articles were included in the meta-analysis. The details of the screening process are described in Figure 1.

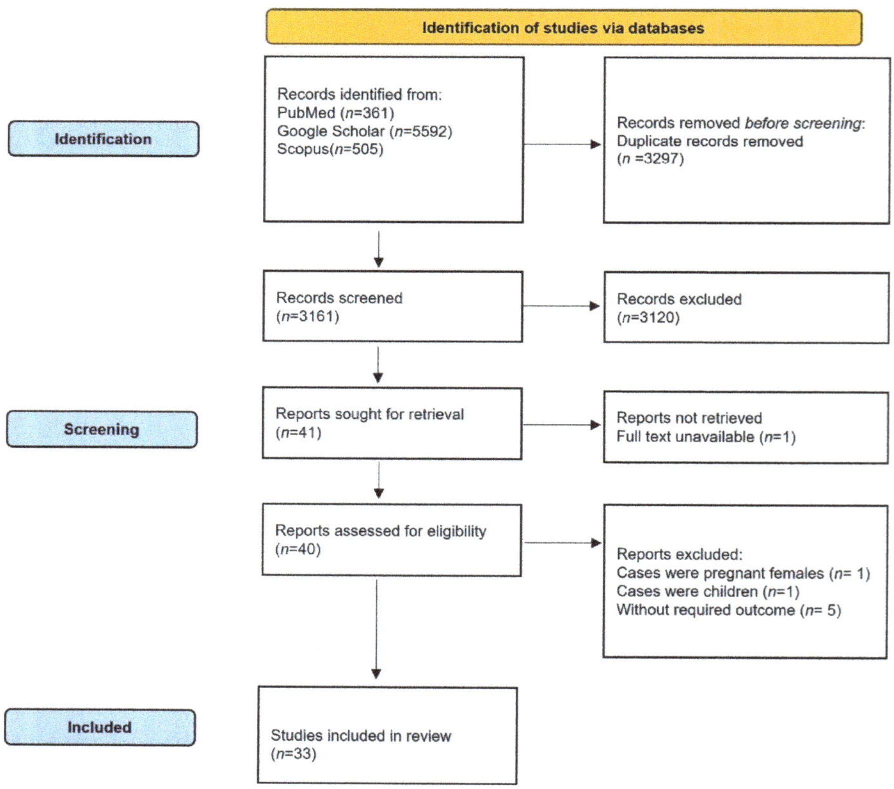

Figure 1. PRISMA 2020 flow diagram of the study selection process.

3.2. Study Characteristics and Quality Assessment

The studies included in this review were published until 2022. All the studies were in English. The studies were conducted in China ($n = 9$), Iran ($n = 4$), Egypt ($n = 3$), Turkey ($n = 3$), Germany ($n = 2$), India ($n = 2$), Pakistan ($n = 2$), Poland ($n = 2$), Taiwan ($n = 2$), Israel ($n = 1$), Korea ($n = 1$), and Vietnam ($n = 1$). One study was a multi-centric study conducted in multiple countries.

Out of the studies selected for review, 18 were cross-sectional studies, 10 were case-control studies, 2 were cohort studies, and 3 were randomized controlled trials. All the studies had cases with CHB or inactive HBV carriers as the study population. One study involved patients with CTP-A cirrhosis and one study had patients with HBsAg seroclearance. Only the relevant and suitable data of CHB patients in each study were included in the meta-analysis.

As per the Downs and Black quality assessment, 8 studies were graded as good quality, 23 as fair and 2 studies were graded as poor. The agreement between the reviewers was calculated using the Cohen's kappa statistic, and showed almost perfect agreement (99.7%) with Cohen's k value 0.927 [17]. The study characteristics of the studies included in the review are described in Table 2.

3.3. Pooled Estimates

Thirty studies reported the mean 25-hydroxy-vitamin D levels. The pooled estimate for 25-hydroxy-vitamin D levels among CHB cases was 21.0568 ng/mL (95% CI: 17.5815–24.5321) (Cochran Q test $p < 0.001$, I2 = 99.4%).

3.4. Meta-Analysis

Meta-analysis was performed to study the difference between 25-hydroxy-vitamin D levels among cases with CHB and healthy controls. The units of 25-hydroxy-vitamin D levels were transformed from nmol/L to ng/mL to maintain uniformity in the results. As described in the forest plot (Figure 2a) The average serum 25-hydroxy-vitamin D levels were significantly lower in CHB patients than in healthy controls and the pooled mean difference was −0.59 ng/mL (−0.82, −0.35) (Cochran Q test $p < 0.001$, I2 = 91.1%). As seen in Figure 2b, the pooled mean difference of vitamin D levels between HBV carriers and healthy controls was −1.32 ng/mL (−2.84, −0.20) (Cochran Q test $p < 0.01$, I2 = 96%).

Table 2. Main characteristics of included studies.

Author & Year	Country	Study Design	Number of Cases	CHB State	Number of Healthy Controls	Vitamin D Estimation Method	Vitamin D in Patients (Mean ± SD) (ng/mL)	Vitamin D in Controls (Mean ± SD) (ng/mL)	Quality Score
Said et al. 2017 [18]	Egypt	retrospective case control	96	Inactive carriers	25	ELISA	13.3 ± 4.1	27 ± 6.76	17
Demir et al. 2013 [19]	Turkey	cross-sectional	35	Inactive carriers	30	radioimmunoassay	7.65 ± 4.19	12.1 ± 7.13	18
Sali et al. 2016 [20]	Iran	retrospective case control	28	Inactive carriers	32	chemiluminescence	32.89 ± 32.5	44.84 ± 34.33	13
Lin et al. 2022 [21]	China	retrospective case control	363	CHB	80	electrochemiluminescence based assay	14.25 ± 7.09	23.6 ± 6.2	22
Zhao et al. 2016 [22]	China	retrospective case control	115	CHB	115	HPLC-TMS	7.83 ± 3.47	9.76 ± 4.36	21
Luo et al. 2022 [23]	China	cross-sectional	898	CHB & MAFLD with CHB	360	electrochemiluminescence based assay	27.2371 ± 9.277	29 ± 9.5	20
Kowerda et al. 2019 [24]	Poland	retrospective case control	58	CHB	9	ELISA	29.6646 ± 5.0069	33.2984 ± 4.2131	15
Sajith et al. 2017 [25]	India	retrospective case control	215	CHB	58	chemiluminescence	24.0223 ± 11.7819	26.8 ± 8.7	16
Mou et al. 2018 [26]	China	retrospective case control	52	CHB	40	ELISA	15.83 ± 3.94	18.34 ± 5.26	16
Chen et al. 2015 [27]	China	cohort	128	CHB	128	electrochemiluminescence based assay	16.88 ± 6.4	20.16 ± 5.5	22
Hoan et al. 2016 [9]	Vietnam	cross-sectional	165	CHB	122	ELISA	21.2 ± 8.9	23.6 ± 9.5	17
Albas et al. 2021 [28]	Turkey	cross-sectional	86	CHB	86	chemiluminescence	10.16 ± 5.1423	11.1895 ± 6.4995	16
Mashaly et al. 2016 [29]	Egypt	cross-sectional	52	CHB	34	ELISA	11.4 ± 6.7	21.1 ± 8.7	19
Naguib et al. 2022 [30]	Egypt	retrospective case control	60	CHB	60	chemiluminescence	19.02 ± 8.42	22.43 ± 8.96	14
Osmani et al. 2021b [31]	Iran	retrospective case control	292	CHB	304	electrochemiluminescence based assay	17.76 ± 5.53	22.07 ± 2.41	20
Thakur et al. 2021 [32]	India	cross-sectional	30	Hep B Cirrhosis	30	electrochemiluminescence based assay	25.4 ± 11	30.4 ± 8.6	16
Mahamid et al. 2013 [33]	Israel	cross-sectional	53	HBsAg seroclearance	-	NA	28.0283 ± 8.0753	-	16

Table 2. *Cont.*

Author & Year	Country	Study Design	Number of Cases	CHB State	Number of Healthy Controls	Vitamin D Estimation Method	Vitamin D in Patients (Mean ± SD) (ng/mL)	Vitamin D in Controls (Mean ± SD) (ng/mL)	Quality Score
Yu R et al. 2018 [34]	China	RCT	560	CHB	-	electrochemiluminescence based assay	29.64 ± 11.29	-	15
Yu R et al. 2015 [35]	China	cross-sectional	242	CHB	-	electrochemiluminescence based assay	33.9 ± 10.67	-	16
Chan H et al. 2015 [36]		RCT	737	CHB	-	chemiluminescence	18.4 ± 7.46	-	16
Gao W et al. 2017 [37]	China	cross-sectional	100	CHB	-	NA	15.3 ± 5.6	-	16
Farnik et al. 2013 [11]	Germany	retrospective case control	203	CHB	-	radioimmunoassay	14.4 ± 7.9	-	16
Osmani et al. 2021a [38]	Iran	cross-sectional	292	CHB	-	electrochemiluminescence based assay	18.4 ± 3.5	-	16
Ko et al. 2016 [39]	Korea	cross-sectional	207	CHB	-	isotope-dilution liquid chromatography-tandem mass spectrometry.	13.4717 ± 7.1565	-	16
Ko et al. 2020 [40]	Taiwan	cross-sectional	60	CHB	-	chemiluminescence	20.9 ± 5.6	-	16
Wong et al. 2014 [41]	China	cohort	426	CHB	-	electrochemiluminescence based assay	24.3 ± 9.4	-	16
Berkan-Kawinska et al. 2015 [42]	Poland	cross-sectional	35	CHB	-	chemiluminescence	17.6 ± -	-	15
Hashemi et al. 2020 [43]	Iran	cross-sectional	281	CHB	-	ELISA	23.69 ± 11.26	-	15
Karim et al. 2021 [44]	Pakistan	cross-sectional	108	CHB	-	chemiluminescence	25.23 ± -	-	10
Kumar et al. 2021 [45]	Pakistan	cross-sectional	93	CHB	-	NA	24.31 ± -	-	10
Motor et al. 2014 [46]	Turkey	cross-sectional	81	Inactive carriers	-	chemiluminescence	52.764 ± 20.03	-	13
Schiefke et al. 2005 [47]	Germany	cross-sectional	13	CHB	-	biochemistry assay	31.2354 ± 13.3896	-	13
Wang et al. 2020 [48]	Taiwan	RCT	196	CHB	-	chemiluminescence	19.8 ± 7.4	-	21

RCT: Randomised Controlled Trial, CHB: Chronic Hepatitis B, MAFLD: Metabolic-associated fatty liver disease, HPLC-TMS: High Performance Liquid Chromatography Tandem Mass Spectrometry method, NA: Not Available.

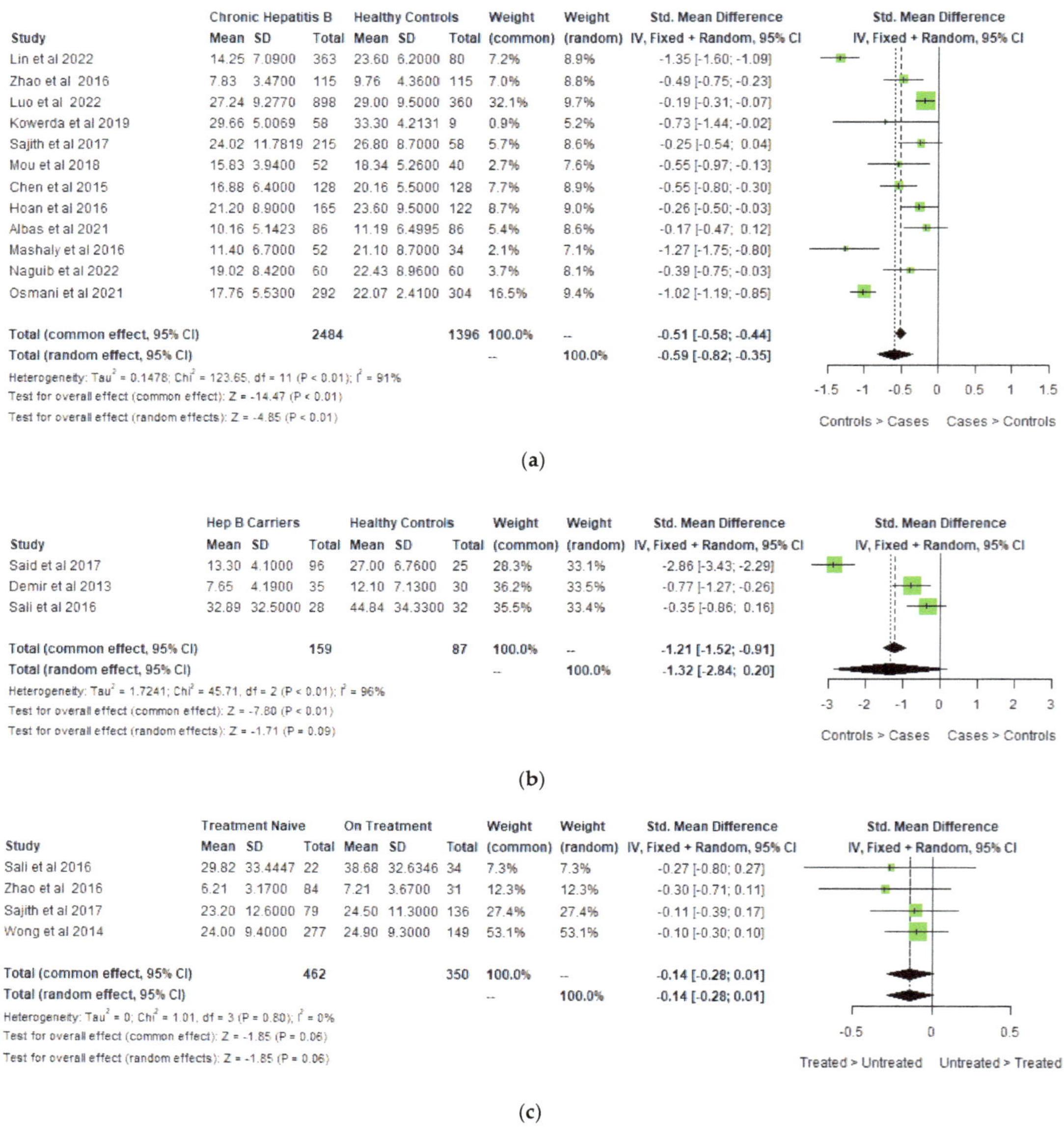

Figure 2. Forest plot of vitamin D levels between (**a**) CHB cases and healthy controls [9,21–31], (**b**) inactive HBV carriers and healthy controls [18–20], (**c**) treatment-naïve and on-treatment Hepatitis B patients [20,22,25,41].

The funnel plot for 25-hydroxy-vitamin D levels in CHB versus healthy controls is asymmetrical, as shown in Figure 3a. The Egger's test, however, suggested no evidence of potential publication bias (p = 0.4143).

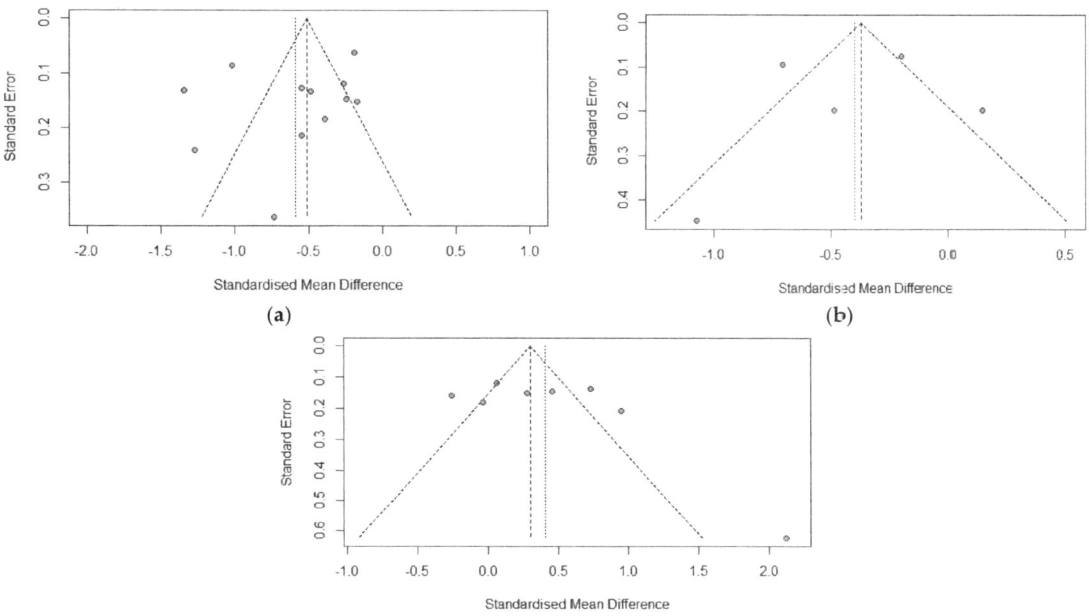

Figure 3. Funnel plot for publication bias analysis for vitamin D levels between (**a**) CHB patients and healthy controls (**b**) HBeAg-positive and negative CHB patients (**c**) CHB patients with or without liver disease.

The differences in 25-hydroxy-vitamin D levels according to HBeAg status, liver cirrhosis/fibrosis, and the effect of antiviral treatment were also analyzed. Five studies studied the differences in 25-hydroxy-vitamin D levels according to HBeAg status and found that 25-hydroxy-vitamin D levels were lower among HBeAg-positive patients as compared to HBeAg-negative patients. The pooled mean difference among the HBeAg-positive and negative groups was −0.4 ng/mL (−0.75, −0.05) (Cochran Q test $p < 0.001$, I2 = 85%). (Figure 4) The funnel plot for 25-hydroxy-vitamin D levels in HBeAg-positive versus negative cases is almost symmetrical with one outlier, as shown in Figure 3b. Five studies studied the 25-hydroxy-vitamin D levels in patients with cirrhosis and no cirrhosis, with a pooled mean difference of −0.48 ng/mL (−0.78, −0.18) (Cochran Q test $p < 0.001$, I2 = 82%). Three studies studied the 25-hydroxy-vitamin D levels in patients with fibrosis and no fibrosis with a pooled mean difference of −0.50 ng/mL (−1.86, 0.87) (Cochran Q test p <0.001, I2 = 85%). (Figure 5) The funnel plot for 25-hydroxy-vitamin D levels in CHB patients with or without liver disease shows studies concentrated around the top of the funnel, with only one outlier, as shown in Figure 3c. Four studies explored the difference in 25-hydroxy-vitamin D levels in treatment-naïve patients and patients on anti-viral treatment with a pooled mean difference of −0.14 ng/mL (−1.86, 0.87) (Cochran Q test $p < 0.001$, I2 = 85%) (Figure 2c).

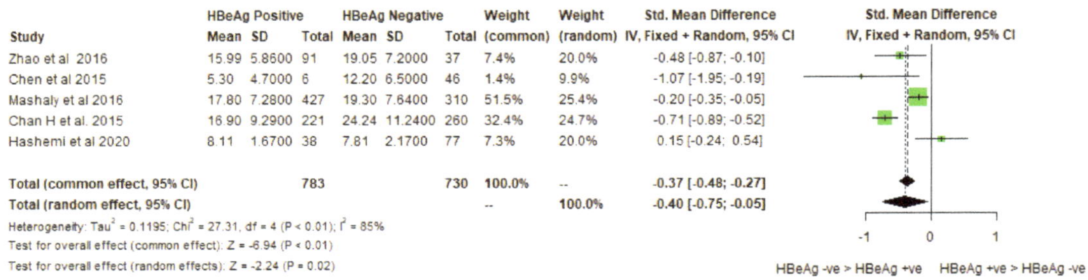

Figure 4. Forest plot of vitamin D levels between HBeAg-positive and HBeAg-negative Chronic Hepatitis B cases [22,27,29,36,43].

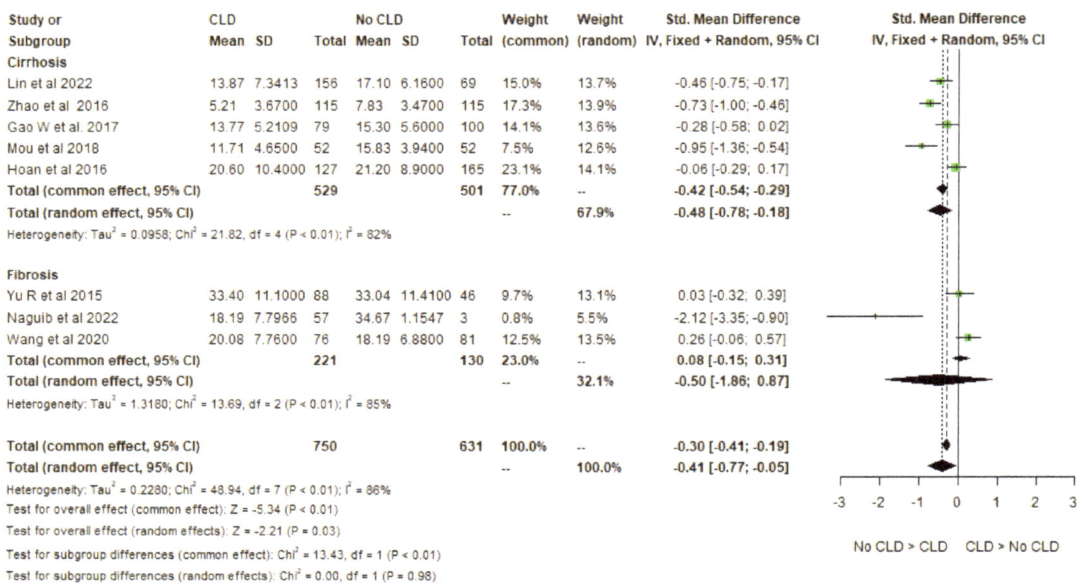

Figure 5. Forest plots of vitamin D levels in CHB patients with absence or presence of cirrhosis [9,21,22,26,37]/fibrosis [30,35,48].

3.5. Sensitivity Analysis

After excluding four studies that were lying outside the funnel plot, the standardized mean difference between CHB cases and controls was 0.3274 ng/mL [0.2104, 0.4443], which was statistically significant with a p value < 0.001 (Cochran Q test $p = 0.1238$, $I^2 = 38.4\%$)

3.6. Meta-Regression

We performed meta-regression analysis to see if latitude, age, male-to-female ratio among cases and control, and type of assay used for detection had any effect on the standardized mean difference of 25-hydroxy-vitamin D levels among CHB and healthy controls. As shown in Table 3, we could not find any significant association between these variables and the standardized mean difference, except that the method of detection had a significant impact in CHB cases ($p = 0.0306$). The scatterplot of latitude, methods, and gender ratio in cases versus controls was plotted as shown in Figures 6–8.

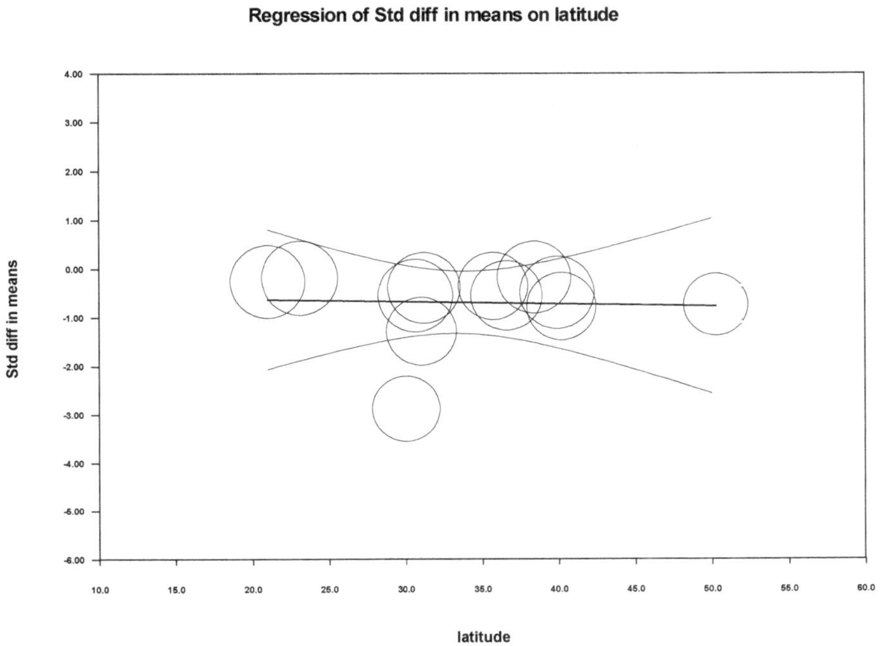

Figure 6. Cases vs. Controls: Scatterplot for latitude.

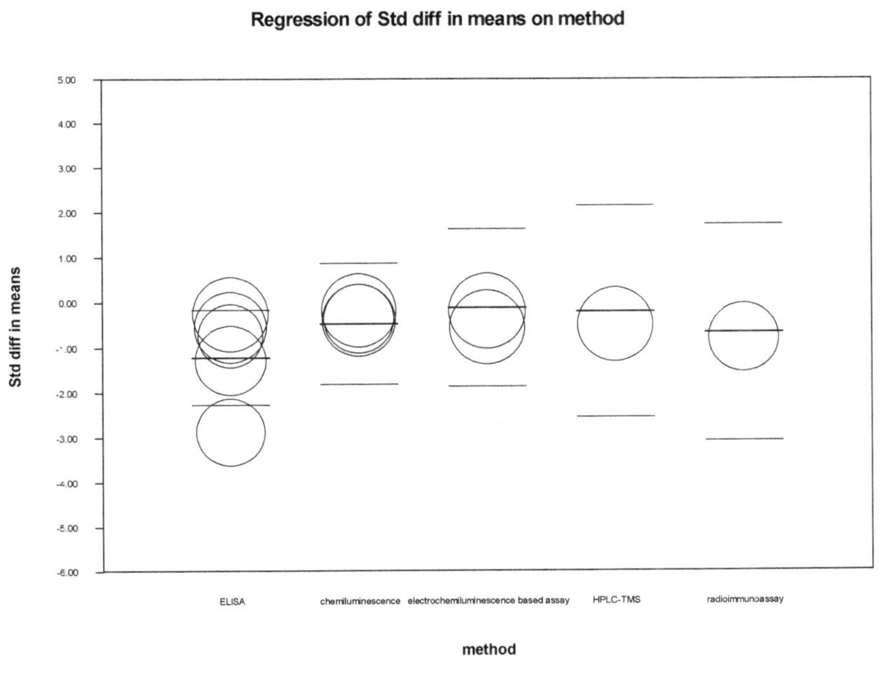

Figure 7. Cases vs. Controls: Scatterplot for method.

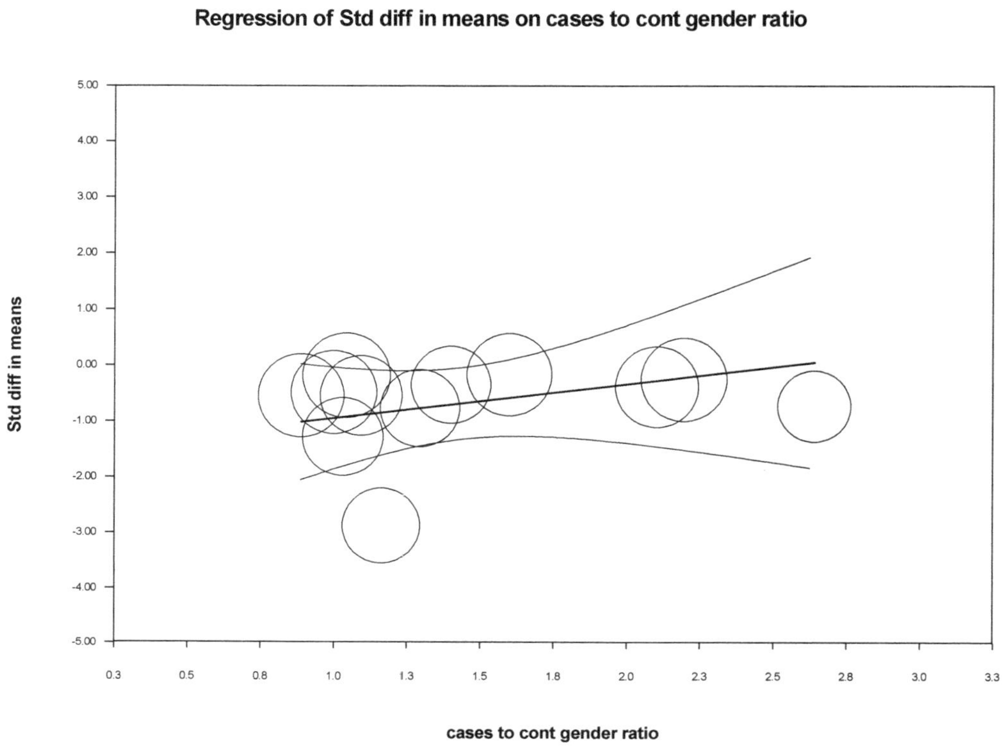

Figure 8. Cases vs. Controls: Scatterplot for Gender Ratio in Cases to Controls.

Table 3. Random effect meta-regression analysis of 25-hydroxy-vitamin D levels.

Covariate	Coefficient	Standard Error	95% Lower	95% Upper	Z-Value	2-Sided *p*-Value	Set
Cases vs. Controls							
Intercept	−1.9311	1.0683	−4.025	0.1628	−1.81	0.0707	
Latitude	−0.0047	0.0273	−0.0581	0.0488	−0.17	0.8635	
Chemiluminescence Method	0.7477	0.4402	−0.1152	1.6105	1.7	0.0894	Q = 6.03, df = 4, *p* = 0.1971
Electrochemiluminescence Method	1.1013	0.5598	0.004	2.1985	1.97	0.0492	
HPLC-TMS Method	1.0181	0.7092	−0.3718	2.4081	1.44	0.1511	
Radioimmunoassay Method	0.5537	0.7118	−0.8414	1.9488	0.78	0.4366	
Cases to Controls Gender Ratio	0.6102	0.3962	−0.1663	1.3867	1.54	0.1235	
All Cases							
Intercept	53.1721	29.3444	−4.3418	110.686	1.81	0.07	
Chemiluminescence Method	−3.2579	10.6526	−24.1366	17.6208	−0.31	0.7597	Q = 13.91, df = 6, *p* = 0.0306
Electrochemiluminescence Method	−9.2847	11.8277	−32.4666	13.8971	−0.78	0.4325	
ELISA Method	−13.5263	10.6162	−34.3337	7.2811	−1.27	0.2026	
HPLC-TMS Method	−19.5501	12.7232	−44.4871	5.3868	−1.54	0.1244	

Table 3. *Cont.*

Covariate	Coefficient	Standard Error	95% Lower	95% Upper	Z-Value	2-Sided p-Value	Set
Isotope-dilution Liquid Chromatography-tandem Mass Spectrometry Method	−15.438	12.1771	−39.3047	8.4288	−1.27	0.2049	
Radioimmunoassay Method	−27.7252	11.4111	−50.0906	−5.3598	−2.43	0.0151	
Latitude	0.2235	0.2937	−0.3521	0.7991	0.76	0.4466	
Age	−0.6979	0.4296	−1.5398	0.1441	−1.62	0.1043	
Male female ratio	0.3166	2.1212	−3.8409	4.474	0.15	0.8814	
All Controls							
Intercept	−71.6418	82.2164	−232.783	89.4993	−0.87	0.3835	
Latitude	0.9468	0.8782	−0.7744	2.668	1.08	0.231	
Electrochemiluminescence Method	−9.1972	16.0905	−40.734	22.3395	−0.57	0.5676	
ELISA Method	−2.4554	8.7522	−19.6094	14.6985	−0.28	0.7791	Q = 3.09, df = 4, $p = 0.5421$
HPLC-TMS Method	−60.6786	43.2949	−145.535	24.1777	−1.4	0.1611	
Radioimmunoassay Method	−8.1006	15.618	−38.7113	22.5101	−0.52	0.604	
Male female ratio	14.047	13.5259	−12.4632	40.5572	1.04	0.299	
Age	1.0935	1.0572	−0.9785	3.1655	1.03	0.3009	

4. Discussion

According to WHO, the current global burden of chronic hepatitis B-infected people is around 296 million people, with around 1.5 million new infections added every year [49]. A total of 33 studies were included in the present analysis, covering a total population of 6360, with 6037 for chronic hepatitis B, and 240 inactive carriers. The majority of the studies used 25-hydroxy-vitamin D to assess serum vitamin D levels in chronic hepatitis B patients. Our results indicate that chronic hepatitis B infection was associated with reduced 25-hydroxy-vitamin D levels. The role of 25-hydroxy-vitamin D in hepatitis B has been implicated by various studies, though its association with disease status has not been analyzed. This meta-analysis has explained the association of serum 25-hydroxy vitamin D levels in different stages of chronic hepatitis B patients. Our study has shown that significantly low levels of 25-hydroxy-vitamin D are associated with chronic hepatitis B patients compared to healthy controls. Furthermore, we found that this association was also found among inactive carriers of hepatitis B infection. In addition, our results also pointed out that serum levels of 25-hydroxy-vitamin D were further low in treatment-naïve patients compared to those on antiviral treatment. Thus, it would be plausible to suggest that lower 25-hydroxy-vitamin D levels can influence the viral outcome, and vice versa, pointing toward a multifaceted crosstalk among host and viral factors. As a causal factor, low serum 25hydroxy-vitamin D levels might influence the viral outcome by affecting appropriate immune response; thus, leading to chronicity. On the other hand, liver inflammation and pathology in hepatitis can compromise the 25 hydroxylation of cholecalciferol in the liver, the key step in vitamin D activation.

A study explored the effect of Vitamin D supplementation on HBV replication but did not find any significant change in HBV DNA levels before and after supplementation. Furthermore, this study studied the effect after supplementation of 2 months, and the long-term effect of supplementation on viral activity remains to be explored [48]. Another study has suggested that impaired liver function in HBV-related cirrhosis could be responsible for insufficient vitamin D hydroxylation and subsequent activation leading to reduced levels in the blood [9]. In line with this, one prospective study has found that serum 25-hydroxy-vitamin D levels negatively correlated with the severity of cirrhosis, with the lowest levels found in decompensated end-stage liver disease [50]. Moreover, as the liver is the principal organ for different transport protein synthesis, reduced production of 25-hydroxy-vitamin D binding protein in chronic hepatitis could further promote insufficient 25-hydroxy-vitamin D levels in circulation. Lastly, ethnic and geo-environmental factors such as location of residence, skin tone, seasonal variations, exposure to sunlight, and nutritional intake can influence the serum 25-hydroxy-vitamin D3 levels

Our next key observation was reduced levels of 25-hydroxy-vitamin D levels in HBe antigen-positive patients compared to HBeAg-negative ones. It highlights that 25-hydroxy-vitamin D levels affect the E antigen secretion viral proliferation in the host. In agreement with this, another study has found that 25-hydroxy-vitamin D levels inversely correlated with HBV DNA load in chronic hepatitis B patients [12].

Although a previous meta-analysis has shown the inverse correlation between 25-hydroxy-vitamin D levels and HBV DNA in chronic hepatitis B patients, no study has ever explored the detailed analysis of all studies estimating 25-hydroxy-vitamin D levels in hepatitis B patients with a different disease or treatment status. Our study has shown that 25-hydroxy-vitamin D levels varied significantly in treatment-naïve CHB patients compared to those on antiviral treatment. This again strengthens the fact that HBV infection in the liver impairs 25-hydroxy-vitamin D metabolism and its levels. Host immunological response is also governed by 25-hydroxy-vitamin D levels. A study on hepatitis B patients on interferon therapy has found that HBV DNA levels decreased more rapidly in patients with a higher level of serum 25-hydroxy-vitamin D. Thus, it signifies that 25-hydroxy-vitamin D influences the cellular immune response to a great extent [51].

Furthermore, results from subgroup analysis among CHB patients with fibrosis and cirrhosis indicated that 25-hydroxy-vitamin D levels positively correlated with fibrosis severity by the common effects model [$p < 0.01$]. A study administering ergo/cholecalciferol in alcoholic liver disease patients has demonstrated that the biological availability of 25-hydroxy-vitamin D positively correlated with patients with mild to moderate liver fibrosis compared with patients with severe liver disease (Child Pugh-C) [52]. The negative correlation between the severity of fibrosis/cirrhosis and 25-hydroxy-vitamin D levels could also be due to the fact that vitamin D, through the activation (VDR) and Calcium-sensing receptor (CaSR), imparts portal hypotensive effect as evident from a study in a rat model [53].

Meta-regression was performed to analyze how common variables such as age, sex, latitude, and the method of 25-hydroxy-vitamin D estimation influenced or showed any association with the serum 25-hydroxy-vitamin D level (intervention effect) in the meta-analysis. The analysis showed no significant association between these variables with the estimation of serum Vitamin D levels in cases vs. healthy control, suggesting these factors were adjusted and nullified.

However, the study has some limitations which should be considered while interpreting the results. The heterogeneity among studies was quite high due to different study types, sample size, and different disease states related to chronic HBV. There was also significant publication bias as evident from our findings. However, sensitivity analysis was performed to check the robustness of the results and models and it was found that after excluding four studies the mean difference remained statistically significant. Lastly, although we tried to incorporate all eligible published studies, there are still chances of missing a few studies from the grey literature.

5. Conclusions

The meta-analysis has covered the most updated and pooled estimate of serum 25-hydroxy-vitamin D levels in altogether and different disease states of chronic hepatitis B patients. Presumably, this is the first systematic review and meta-analysis which has identified major differences in serum 25-hydroxy-vitamin D levels correlating with disease activity such as HBeAg status and severity in terms of fibrosis and cirrhosis. Notwithstanding the wide heterogeneity among the included studies, our analysis strongly suggests that serum 25-hydroxy-vitamin D levels are associated with disease activity and pathobiology, although the exact nature of the cause-effect relationship cannot be discerned from this study. Future research is necessary to conduct in this area to validate the therapeutic and preventive role of vitamin D against chronic hepatitis B and related liver diseases.

Thus, this detailed meta-analysis provides corroborative evidence about the role of vitamin D in hepatitis B-related diseases and suggests that monitoring 25-hydroxy-vitamin D status and supplementation of vitamin D in chronic hepatitis B patients to prevent long-term complications should be addressed in future well-designed study cohorts.

Author Contributions: Conceptualization, A.B. and N.K.; methodology, S.A. and A.B.; software, N.K. and S.A.; validation, N.K., S.A. and A.B.; formal analysis, P.S.; resources, N.K., S.A. and A.B.; data curation, S.A.; writing—original draft preparation, A.B.; writing—review and editing, N.K., M.M.; visualization, S.A., P.S.; supervision, A.B.; project administration, P.S. and M.M. All authors have read and agreed to the published version of the manuscript.

Funding: This research received no external funding.

Institutional Review Board Statement: Not applicable.

Informed Consent Statement: Not applicable.

Data Availability Statement: All relevant data were included in the paper.

Conflicts of Interest: The authors declare no conflict of interest.

References

1. Tang, L.S.Y.; Covert, E.; Wilson, E.; Kottilil, S. Chronic Hepatitis B Infection: A Review. *JAMA* **2018**, *319*, 1802–1813. [CrossRef]
2. Peng, C.-Y.; Chien, R.-N.; Liaw, Y.-F. Hepatitis B Virus-Related Decompensated Liver Cirrhosis: Benefits of Antiviral Therapy. *J. Hepatol.* **2012**, *57*, 442–450. [CrossRef] [PubMed]
3. Binkley, N.; Ramamurthy, R.; Krueger, D. Low Vitamin D Status: Definition, Prevalence, Consequences, and Correction. *Endocrinol. Metab. Clin. N. Am.* **2010**, *39*, 287–301, table of contents. [CrossRef] [PubMed]
4. Nair, R.; Maseeh, A. Vitamin D: The "Sunshine" Vitamin. *J Pharm. Pharm.* **2012**, *3*, 118–126. [CrossRef]
5. Ahluwalia, S.; Choudhary, D.; Tyagi, P.; Kumar, V.; Vivekanandan, P. Vitamin D Signaling Inhibits HBV Activity by Directly Targeting the HBV Core Promoter. *J. Biol. Chem.* **2021**, *297*, 101233. [CrossRef] [PubMed]
6. Yin, K.; Agrawal, D.K. Vitamin D and Inflammatory Diseases. *J. Inflamm. Res.* **2014**, *7*, 69–87. [CrossRef]
7. Ratnam, D.; Visvanathan, K. New Concepts in the Immunopathogenesis of Chronic Hepatitis B: The Importance of the Innate Immune Response. *Hepatol. Int.* **2008**, *2*, 12–18. [CrossRef]
8. Kondo, Y.; Ueno, Y.; Shimosegawa, T. Toll-like Receptors Signaling Contributes to Immunopathogenesis of HBV Infection. *Gastroenterol. Res. Pract.* **2011**, *2011*, 810939. [CrossRef]
9. Hoan, N.X.; Khuyen, N.; Binh, M.T.; Giang, D.P.; Van Tong, H.; Hoan, P.Q.; Trung, N.T.; Anh, D.T.; Toan, N.L.; Meyer, C.G.; et al. Association of Vitamin D Deficiency with Hepatitis B Virus—Related Liver Diseases. *BMC Infect. Dis.* **2016**, *16*, 507. [CrossRef]
10. Nimer, A.; Mouch, A. Vitamin D Improves Viral Response in Hepatitis C Genotype 2-3 Naïve Patients. *World J. Gastroenterol.* **2012**, *18*, 800–805. [CrossRef]
11. Farnik, H.; Bojunga, J.; Berger, A.; Allwinn, R.; Waidmann, O.; Kronenberger, B.; Keppler, O.T.; Zeuzem, S.; Sarrazin, C.; Lange, C.M. Low Vitamin D Serum Concentration Is Associated with High Levels of Hepatitis B Virus Replication in Chronically Infected Patients. *Hepatology* **2013**, *58*, 1270–1276. [CrossRef]
12. Hu, Y.-C.; Wang, W.-W.; Jiang, W.-Y.; Li, C.-Q.; Guo, J.-C.; Xun, Y.-H. Low Vitamin D Levels Are Associated with High Viral Loads in Patients with Chronic Hepatitis B: A Systematic Review and Meta-Analysis. *BMC Gastroenterol.* **2019**, *19*, 84. [CrossRef]
13. Page, M.J.; McKenzie, J.E.; Bossuyt, P.M.; Boutron, I.; Hoffmann, T.C.; Mulrow, C.D.; Shamseer, L.; Tetzlaff, J.M.; Akl, E.A.; Brennan, S.E.; et al. The PRISMA 2020 Statement: An Updated Guideline for Reporting Systematic Reviews. *BMJ* **2021**, *372*, n71. [CrossRef]
14. Ouzzani, M.; Hammady, H.; Fedorowicz, Z.; Elmagarmid, A. Rayyan—A Web and Mobile App for Systematic Reviews. *Syst. Rev.* **2016**, *5*, 210. [CrossRef]
15. Downs, S.H.; Black, N. The Feasibility of Creating a Checklist for the Assessment of the Methodological Quality Both of Randomised and Non-Randomised Studies of Health Care Interventions. *J. Epidemiol. Community Health* **1998**, *52*, 377–384. [CrossRef]
16. Faul, F.; Erdfelder, E.; Buchner, A.; Lang, A.-G. Statistical Power Analyses Using G*Power 3.1: Tests for Correlation and Regression Analyses. *Behav. Res. Methods* **2009**, *41*, 1149–1160. [CrossRef]
17. Landis, J.R.; Koch, G.G. The Measurement of Observer Agreement for Categorical Data. *Biometrics* **1977**, *33*, 159–174. [CrossRef]
18. Said, E.; Agawy, W.E.; Ahmed, R.; Hassany, M.; Ahmed, A.; Fouad, H.; Baiumy, H. Serum Vitamin D Levels in Treatment-Naïve Chronic Hepatitis B Patients. *J. Transl. Intern. Med.* **1977**, *5*, 230–234. [CrossRef]

19. Demir, C.; Demir, M. Vitamin D Levels in Patients with Chronic Hepatitis B Virus Infection and Naturally Immunized Individuals. *Intern. Med. Inside* **2013**, *4*, 2343–6549. [CrossRef]
20. Sali, S.; Tavakolpour, S.; Farkhondemehr, B. Comparison of Vitamin D Levels in Naive, Treated, and Inactive Carriers with Chronic Hepatitis B Virus. *J. Clin. Transl. Hepatol.* **2016**, *4*, 306–309. [CrossRef]
21. Lin, S.; Wang, W.; Shi, L.; Yang, X.; Chen, Y.; Liu, X.; Li, J.; Ye, F.; An, X.; Zhang, X. Severe Vitamin D Deficiency Is Strongly Associated with Liver Dysfunction and Disease Severity in Hepatitis B Virus Related Cirrhosis and Liver Failure Patients. *J. Nutr. Sci. Vitaminol.* **2022**, *68*, 16–22. [CrossRef] [PubMed]
22. Zhao, X.Y.; Li, J.; Wang, J.H.; Habib, S.; Wei, W.; Sun, S.J.; Strobel, H.W.; Jia, J.D. Vitamin, D. Serum Level Is Associated with Child-Pugh Score and Metabolic Enzyme Imbalances, but Not Viral Load in Chronic Hepatitis B Patients. *Medicine* **2016**, *95*, e3926. [CrossRef] [PubMed]
23. Luo, L.; Ye, J.; Shao, C.; Lin, Y.; Sun, Y.; Feng, S.; Wang, W.; Zhong, B. Vitamin D Status Presents Different Relationships with Severity in Metabolic-Associated Fatty Liver Disease Patients with or without Hepatitis B Infection. *Nutrients* **2022**, *14*, 2114. [CrossRef] [PubMed]
24. Parfieniuk-Kowerda, A.; Świderska, M.; Rogalska, M.; Maciaszek, M.; Jaroszewicz, J.; Flisiak, R. Chronic Hepatitis B Virus Infection Is Associated with Decreased Serum 25(OH)D Concentration in Non-Cirrhotic Patients. *Clin. Exp. Hepatol.* **2019**, *5*, 75–80. [CrossRef]
25. Sajith, K.G.; Kapoor, N.; Shetty, S.; Goel, A.; Zachariah, U.; Eapen, C.E.; Paul, T.V. Bone Health and Impact of Tenofovir Treatment in Men with Hepatitis-B Related Chronic Liver Disease. *J. Clin. Exp. Hepatol.* **2018**, *8*, 23–27. [CrossRef]
26. Mou, H.; Yang, F.; Zhou, J.; Bao, C. Correlation of Liver Function with Intestinal Flora, Vitamin Deficiency and IL-17A in Patients with Liver Cirrhosis. *Exp. Ther. Med.* **2018**, *16*, 4082–4088. [CrossRef]
27. Chen, E.Q.; Bai, L.; Zhou, T.Y.; Fe, M.; Zhang, D.M.; Tang, H. Sustained Suppression of Viral Replication in Improving Vitamin D Serum Concentrations in Patients with Chronic Hepatitis B. *Sci. Rep.* **2015**, *5*, 15441. [CrossRef]
28. Albas, S.; Koc, E.; Nemli, S.; Demirdal, T.; Soyoz, M.; Aksun, S.; Sozmen, M.K.; Avsar, C.; Gurbu, B.C. Vitamin D Levels and Vitamin D Receptor (VDR) Gene Polymorphisms in Inactive Hepatitis B Virus Carriers. *Parameters* **2021**, *30*, 393–398.
29. Mashaly, M.; Sayed, E.E.; Shaker, G.A.; Anwar, R.; Abbas, N.F.; Zakaria, S.; Barakat, E.A.M.E. Occult and Chronic Hepatitis B Infection: Relation of Viral Load to Serum Level of 25 Hydroxy Vitamin D. *Int. J. Curr. Microbiol. App. Sci.* **2016**, *5*, 660–669. [CrossRef]
30. Naguib, R.; Fayed, A.; Abdeen, N.; Naguib, H. Association of Serum 25-Hydroxyvitamin D3 Levels and Insulin Resistance with Viral Load and Degree of Liver Fibrosis in Egyptian Chronic HBV Patients: A Case Control Study. *Clin. Exp. Hepatol.* **2022**, *8*, 14–20. [CrossRef]
31. Osmanï, F.; Ziaee, M. The Importance of Vitamin D Deficiency as a Potential Marker Among Chronic Hepatitis B Patients. *Viral Hepat. J.* **2021**, *27*, 74–79. [CrossRef]
32. Thakur, P.; Cherian, K.E.; Kapoor, N.; Rebekah, G.; Goel, A.; Zachariah, U.; Eapen, C.E.; Thomas, N.; Paul, T.V. Proximal Hip Geometry, Trabecular Bone Score, Bone Mineral Density and Bone Mineral Parameters in Patients With Cryptogenic and Hepatitis B Related Cirrhosis- A Study From the Indian Subcontinent. *J. Clin. Densitom. Off. J. Int. Soc. Clin. Densitom.* **2022**, *25*, 97–104. [CrossRef]
33. Mahamid, M.; Nseir, W.; Abu Elhija, O.; Shteingart, S.; Mahamid, A.; Smamra, M.; Koslowsky, B. Normal Vitamin D Levels Are Associated with Spontaneous Hepatitis B Surface Antigen Seroclearance. *World J. Hepatol.* **2013**, *5*, 328–331. [CrossRef]
34. Yu, R.; Tan, D.; Ning, Q.; Niu, J.; Bai, X.; Chen, S.; Cheng, J.; Yu, Y.; Wang, H.; Xu, M.; et al. Association of Baseline Vitamin D Level with Genetic Determinants and Virologic Response in Patients with Chronic Hepatitis B. *Hepatol. Res. Off. J. Jpn. Soc. Hepatol.* **2018**, *48*, E213–E221. [CrossRef]
35. Yu, R.; Sun, J.; Zheng, Z.; Chen, J.; Fan, R.; Liang, X.; Zhu, Y.; Liu, Y.; Shen, S.; Hou, J. Association between Vitamin D Level and Viral Load or Fibrosis Stage in Chronic Hepatitis B Patients from Southern China. *J. Gastroenterol. Hepatol.* **2015**, *30*, 566–574 [CrossRef]
36. Chan, H.L.; Elkhashab, M.; Trinh, H.; Tak, W.Y.; Ma, X.; Chuang, W.L.; Kim, Y.J.; Martins, E.B.; Lin, L.; Dinh, P.; et al. Association of Baseline Vitamin D Levels with Clinical Parameters and Treatment Outcomes in Chronic Hepatitis B. *J. Hepatol.* **2015**, *63*, 1086–1092. [CrossRef]
37. Gao, W.; Wang, R.; Wang, X.; Wu, H.; Wang, Y.; Lu, X.; Li, L.; Zheng, J.; Li, W. Vitamin D Serum Levels and Receptor Genetic Polymorphisms Are Associated with Hepatitis B Virus and HIV Infections and IFN-λ Levels. *Biomark. Med.* **2017**, *11*, 733–740. [CrossRef]
38. Osmani, F.; Azarkar, G. Fitting Logistic Regression Models to Assess Vitamin D Deficiency with Clinical Parameters in Chronic Hepatitis B Patients. *Infect. Dis. Model.* **2021**, *6*, 612–617. [CrossRef]
39. Ko, B.J.; Kim, Y.S.; Kim, S.G.; Park, J.H.; Lee, S.H.; Jeong, S.W.; Jang, J.Y.; Kim, H.S.; Kim, B.S.; Kim, S.M.; et al. Relationship between 25-Hydroxyvitamin D Levels and Liver Fibrosis as Assessed by Transient Elastography in Patients with Chronic Liver Disease. *Gut Liver* **2016**, *10*, 818–825. [CrossRef]
40. Ko, W.S.; Yang, Y.P.; Shen, F.P.; Wu, M.C.; Shih, C.J.; Lu, M.C.; Yan, Y.H.; Chiou, Y.L. The Study of Correlation Between Serum Vitamin D(3) Concentrations and HBV DNA Levels and Immune Response in Chronic Hepatitis Patients. *Nutrients* **2020**, *12*, 1114. [CrossRef]

41. Wong, G.L.; Chan, H.L.; Chan, H.Y.; Tse, C.H.; Chim, A.M.; Lo, A.O.; Wong, V.W. Adverse Effects of Vitamin D Deficiency on Outcomes of Patients with Chronic Hepatitis B. *Clin. Gastroenterol. Hepatol. Off. Clin. Pract. J. Am. Gastroenterol. Assoc.* **2015**, *13*, 783–790.e1. [CrossRef] [PubMed]
42. Berkan-Kawińska, A.; Koślińska-Berkan, E.; Piekarska, A. The Prevalence and Severity of 25-(OH)-Vitamin D Insufficiency in HCV Infected and in HBV Infected Patients: A Prospective Study. *Clin. Exp. Hepatol.* **2015**, *1*, 5–11. [CrossRef]
43. Hashemi, S.J.; Parsi, A.; Hajiani, E.; Masjedizadeh, A.; Shayesteh, A. Relationship Between 25-HydroxyVitamin D Level and Liver Stiffness in Patients with Chronic Hepatitis B Using Transient Elastography. *Hepat. Mon.* **2020**, *20*, e100891. [CrossRef]
44. Karim, A.; Memon, S.H.; Ahmed, J.; Soomro, A.K.; Manan, A. Prevalence Of Vitamin-D Deficiency Among The Patients Of Hbv And Hcv Relatedchronic Liver Disease. *J. Peoples Univ. Med. Health Sci.* **2021**, *11*, 67–72. [CrossRef]
45. Kumar, D.; Khan, M.U.; Kashif, S.M.; Shaikh, M.A.; Nawaz, Z.; Shaikh, U. Association of Vitamin D Deficiency with Hepatitis B and C Virus Infection. *J. Pharm. Res. Int.* **2021**, *33*, 10–14. [CrossRef]
46. Motor, S.; Koksaldi-Motor, V.; Dokuyucu, R.; Ustun, I.; Evirgen, O.; Yilmaz, N.; Onlen, Y.; Gokce, C. Investigation of Vitamin D Levels in Patients with Inactive Hepatitis B Virus Carrier. *Acta Med. Mediterr.* **2014**, *30*, 793–796.
47. Schiefke, I.; Fach, A.; Wiedmann, M.; Aretin, A.-V.; Schenker, E.; Borte, G.; Wiese, M.; Moessner, J. Reduced Bone Mineral Density and Altered Bone Turnover Markers in Patients with Non-Cirrhotic Chronic Hepatitis B or C Infection. *World J. Gastroenterol.* **2005**, *11*, 1843–1847. [CrossRef]
48. Wang, C.-C.; Tzeng, I.-S.; Su, W.-C.; Li, C.-H.; Lin, H.H.; Yang, C.-C.; Kao, J.-H. The Association of Vitamin D with Hepatitis B Virus Replication: Bystander Rather than Offender. *J. Med. Assoc.* **2020**, *119*, 1634–1641. [CrossRef]
49. WHO Hepatitis B Factsheet. Available online: https://www.who.int/news-room/fact-sheets/detail/hepatitis-b (accessed on 25 October 2022).
50. Finkelmeier, F.; Kronenberger, B.; Zeuzem, S.; Piiper, A.; Waidmann, O. Low 25-Hydroxyvitamin D Levels Are Associated with Infections and Mortality in Patients with Cirrhosis. *PLoS ONE* **2015**, *10*, e0132119. [CrossRef]
51. He, L.-J.; Zhang, H.-P.; Li, H.-J.; Wang, J.; Chang, D.-D. Effect of Serum Vitamin D Levels on Cellular Immunity and Antiviral Effects in Chronic Hepatitis B Patients. *Clin. Lab.* **2016**, *62*, 1933–1939. [CrossRef]
52. Malham, M.; Peter Jørgensen, S.; Lauridsen, A.L.; Ott, P.; Glerup, H.; Dahlerup, J.F. The Effect of a Single Oral Megadose of Vitamin D Provided as Either Ergocalciferol (D_2) or Cholecalciferol (D_3) in Alcoholic Liver Cirrhosis. *Eur. J. Gastroenterol. Hepatol.* **2012**, *24*, 172–178. [CrossRef]
53. Lee, P.-C.; Yang, Y.-Y.; Lee, W.-P.; Lee, K.-C.; Hsieh, Y.-C.; Lee, T.-Y.; Lin, H.-C. Comparative Portal Hypotensive Effects as Propranolol of Vitamin D3 Treatment by Decreasing Intrahepatic Resistance in Cirrhotic Rats. *J. Gastroenterol. Hepatol.* **2015**, *30*, 628–637. [CrossRef]

biomedicines

MDPI

Article

Frequency of Vitamin D Receptor Gene Polymorphisms in a Population with a very High Prevalence of Vitamin D Deficiency, Obesity, Diabetes and Hypertension

Salah Gariballa *, Ghada S. M. Al-Bluwi and Javed Yasin

Internal Medicine, College of Medicine and Health Sciences, United Arab Emirates University, Al-Ain P.O. Box 15551, United Arab Emirates; ghadabluwi@uaeu.ac.ae (G.S.M.A.-B.); javed.yasin@uaeu.ac.ae (J.Y.)
* Correspondence: s.gariballa@uaeu.ac.ae; Tel.: +971-37-137-659; Fax: +971-37-672-995

Abstract: Background: Although vitamin D levels and underlying vitamin D receptor (VDR) genetic polymorphisms have been linked to many common diseases including obesity, the association remains unclear. There is also co-existence of pathologically high proportions of obesity and vitamin D deficiency conditions in our UAE society. We therefore aimed to determine the genotypes and allele percentage frequency distribution of four polymorphisms—FokI, BsmI, ApaI and TaqI—in the VDR gene in healthy Emirati individuals and their association with vitamin D levels and chronic conditions including diabetes mellitus, hypertension and obesity. Methods: 277 participants who were part of a randomized controlled trial had their assessment that included clinical and anthropometric data. Whole blood samples were taken for measurements of vitamin D [25(OH) D], four vitamin D receptor gene polymorphism SNPs, including BsmI, FokI, TaqI and ApaI, metabolic and inflammatory markers and related biochemical variables. Multiple logistic regression analysis was used to assess the influence of vitamin D receptor gene SNPs on vitamin D status after adjusting for clinical parameters known to influence vitamin D status in the study population. Results: Overall, 277 participants with a mean ($\pm$SD) age of 41 $\pm$ 12, 204 (74%) of them being female, were included in the study. There were statistically significant differences in vitamin D concentrations between different genotypes of the four VDR gene polymorphisms ($p < 0.05$). There were, however, no statistically significant differences in vitamin D concentrations between subjects with and those without the four VDR gene polymorphisms genotype and alleles except for AA and AG and allele G in Apal SNP ($p < 0.05$). Multivariate analysis revealed no significant independent associations between vitamin D status and the four VDR gene polymorphisms after adjusting for dietary intake, physical activity, sun exposure, smoking and body mass index. In addition, no significant differences were found in the frequency of the genotypes and alleles of the four VDR genes among patients with obesity, diabetes and hypertension compared to those without these medical conditions. Conclusions: Although we found statistically significant differences in vitamin concentrations between different genotypes of the four VDR gene polymorphisms, multivariate analysis revealed no association after adjusting for clinical parameters known to influence vitamin D status. Furthermore, no association was found between obesity and related pathologies and the four VDR gene polymorphisms.

Keywords: vitamin D; VDR receptor; obesity; diabetes; hypertension

1. Introduction

Vitamin D an active hormone is essential for both bone and muscle health. There is also mounting evidence that vitamin D can convey other health benefits because of the discovery that most cells and tissues in the human body contain vitamin D receptors [1]. Vitamin D elicits its biological action through binding to the vitamin D receptor (VDR) [1]. A number of loss-of-function mutations of the VDR gene have been identified which result in hereditary vitamin-D-resistant rickets [2]. In addition, several subtle allelic polymorphisms which have been linked to diseases such as osteoporosis and metabolic bone disorders have been identified in the VDR gene [3]. However, the health impacts of these single-nucleotide polymorphisms (SNPs) are not well understood [4]. More SNPs are expected to be identified in future because of the large size of the VDR gene and a number of those already described have been found to be associated with some chronic diseases, including multiple sclerosis, rheumatoid arthritis, and some autoimmune and metabolic diseases [5–8]. In addition, the relationship between vitamin D and its VDR receptor polymorphism to obesity and related pathologies have been described [9,10]. Both obesity and vitamin D deficiency are highly prevalent in UAE citizens; however, their adverse and additive health implications are not yet clear. Recent evidence points to a role of vitamin D deficiency as a mediator for increased inflammation associated with obesity and related diabetes [11]. Furthermore, several observational studies have raised the possibility of the role of vitamin D in the development of type 2 diabetes and noted that low plasma 25(OH) D concentrations might be a mediator between obesity and increased risk of diabetes [12]. Furthermore, recent evidence also shows that vitamin D metabolism, storage, and action are affected by adiposity and that obese individuals need higher vitamin D doses compared to normal-weight individuals [13].

The BsmI, FokI, TaqI and ApaI single-nucleotide polymorphisms in the VDR gene are the most common and frequently studied [14,15]. Although numerous studies have reported the frequency of VDR gene SNPs and their associations with different diseases in different ethnic groups (mostly Caucasian populations), very few studies have been conducted in the Middle East, namely the UAE. Moreover, many of those studies conducted in the Middle East suffer from methodological flaws, including selection bias and small sample sizes [13–17]. Mutations of the VDR gene and allelic polymorphisms may influence disease susceptibility and responses to levels of circulating vitamin D, especially in areas where D deficiency is highly prevalent, such as in Asian and Middle Eastern populations. However, results of the studies which have been published so far on the relationship between low vitamin D status and its health implications in non- Caucasian populations yielded conflicting results [18,19]. For example, a cross-sectional study evaluated whether Pakistanis living in the city of Oslo have increased bone turnover compared with ethnic Norwegians due to their high prevalence of vitamin D deficiency [18]. The results revealed only minor ethnic differences in bone turnover, despite a striking difference in prevalence of secondary hyperparathyroidism with no differences observed in bone mineral density between the two groups [18]. Another study from the UK on vitamin D status and markers of bone turnover in Caucasian and South Asian postmenopausal women found that although South Asian women had significantly higher serum parathyroid hormones and lower 25(OH) D concentrations, there were no significant differences between the two groups for biochemical markers of bone turnover [19]. An interesting speculative explanation for this paradox is that altered metabolism of vitamin D due to ethnic and/or genetic differences in the Asian women may protect their skeletons from bone loss [20–22]. This is clearly an area for research. In summary, vitamin D deficiency and obesity and related pathologies are highly prevalent in UAE citizens, but their combined health implications are not yet clear. Mutations of the VDR gene and the presence of SNPs may account for variation in individual responses to levels of circulating vitamin D. The overall objective of this study is to investigate the genotypes and allele percentage frequency distribution of four common polymorphisms, FokI, BsmI, ApaI and TaqI, in the VDR gene in healthy Emirati individuals and their association with Vitamin D levels and chronic conditions, including diabetes mellitus, hypertension and obesity.

2. Materials and Methods

Details of the subjects' recruitment and methods have been published previously [23]. Briefly, healthy community free-living Emirati (UAE citizens) and expatriate Arabian men and women from other countries in the Middle East who were part of a randomized, double-blind placebo-controlled placebo intervention trial were included in this study. Apparently healthy subjects were recruited by local advertisement, from community health centers and from hospital out-patients in Al Ain city, which has two main teaching hospitals serving a total population of 400,000. Individuals with renal disease or stones, hypercalcemia, on calcium and/or vitamin D supplementation, bisphosphonates, steroid medications, hormones or diuretics or unable to give informed written consent were excluded. Following informed written consent, the eligible subjects' blood samples were taken for measurements of vitamin D, vitamin D receptor gene polymorphism SNPs and inflammatory and metabolic risk markers. Clinical assessment that included demographic and baseline characteristics, general and self-rated health and physical activity were performed at baseline. Information on other important variables likely to influence vitamin D status was also collected. Al Ain Medical District Human research ethics committee approved the study protocol and consent to participate form (Reference number: AAMDHREC protocol no 14/15). Written consent obtained from all patients recruited to this study.

2.1. *The Measurements*

Face-to-face questionnaire data were collected on lifestyle and health factors that are of interest in this study of the health implications of vitamin D deficiency in UAE citizens. A common set of questions on education and socio-economic status, current and past occupation, history of previous illness or surgical operations, tobacco smoking, consumption of beverages, physical activity, use of herbal medicine, vitamin supplements and exogenous hormones for contraception and postmenopausal replacement therapy were collected. Anthropometric data including body weight, height and body mass index (BMI) were measured using the Tanita body composition analyzer. Using WHO sex-adjusted cut-of-points for BMI, subjects with BMI = 18–25 were classified as normal, BMI = 25.1–29.9 as overweight and those with BMI 30 as obese.

2.2. *Biochemical and Urine Analysis*

Patients provided a fasting morning blood sample. Biochemical analysis of 25(OH) D was measured using a fully automated COBAS e411 analyzer that uses a patented Electro Chemiluminescence (ECL) technology for immunoassay analysis from ROCHE diagnostics, Mannheim, Germany.

2.3. *DNA Preparation and VDR SNPs Genotyping Analysis*

Genomic DNA was extracted from whole blood collected from study participants using a QIAamp DNA Mini Kit (QIAGEN, Valencia, CA, USA) according to the manufacturer's instructions. The extracted genomic DNA was analyzed using agarose gel electrophoresis, quantitatively determined by spectrophotometry, and stored at 80 °C until use. The four VDR SNPs (BsmI, FokI, TaqI and ApaI) were evaluated using the TaqMan SNP genotyping assay, which consists of a predesigned mix of unlabeled polymerase chain reaction (PCR) primers and the TaqMan® minor groove binding group (MGB) probe (FAM™ and VIC® dye-labeled). All TaqMan SNP genotyping assays are designed to work with the TaqMan® Universal PCR Master Mix, which contains DNA polymerase, dNTPs and optimized mix components and uses the same thermal conditions (Applied Biosystems, Foster City, CA, USA). The PCR consisted of a hot start at 95 °C for 10 min followed by 40 cycles of 94 °C for 15 s and 60 °C for 1 min. Fluorescence detection was performed at 60 °C. All assays were performed in 10–25 µL reactions, using TaqMan Genotyping Master Mix on 96-well plates using Genetic Analyzer (Applied Biosystems, Foster City, CA, USA) according to the manufacturer's instructions.

2.4. *Statistical Power of the Study*

The sample size of 277 would allow the detection of a true mean difference in plasma 25(OH) D of 2.0 (ng/mL) [given the within group SD 8 (ng/mL)] between subjects with and without the VDR gene polymorphism genotype. This sample size would also allow the detection of a 10% difference in the genotypes or allele frequency distribution between subjects with and without obesity (assuming the prevalence is 5% in non-obese subjects and 15% in obese subjects) with 80% power and a type 1 error probability of $\leq$0.05.

2.5. Statistical Analysis

Statistical analysis was performed with SPSS software, version 25.0 (SPSS Inc., Chicago, IL, USA). One-way ANOVA or the nonparametric Kruskal-Wallis H test was used to test for between-group differences, and a p value < 0.05 was considered significant. Multiple regression analyses were used to assess the independent association between vitamin D status (deficiency vs. insufficiency/optimal) and the four VDR genes after adjusting for other clinical parameters including, age, gender, dietary intake, physical activity, sun exposure, smoking and body mass index. The vitamin D status (deficiency vs. insufficiency/optimal) is based on biochemical normative values accepted by many international societies [vitamin D deficiency (<20 ng/mL), insufficiency (20–32 ng/mL) and optimal concentrations (>32 ng/mL)] [17].

3. Results

3.1. Baseline Charachteristics

Overall, 277 participants with a mean (±SD) age of 41 ± 12, 204 (74%) being female, were included in the study. Among the 277 subjects recruited, 46 (17%) had type 2 diabetes and 41 (15%) had hypertension. Using WHO cut-of-points for BMI, 65 (24%) subjects had normal BMI, 93 (34%) were overweight and 108 (39%) were obese at baseline (Table 1).

Table 1. Baseline clinical, anthropometric and metabolic characteristics of the study population.

Variables	Baseline Clinical, Anthropometric and Metabolic Characteristics of the Study Population	Mean (SD) Unless Otherwise Stated (n = 277)
Age (years)		41 (12)
Sex, female n (%)		204 (74)
Smoking n (%)		
	Current	35 (13)
	Ex-smoker	12 (4)
	Never smoked	225 (81)
Body mass index (BMI) n (%)		
	Normal weight (BMI 18.5–25)	65 (24)
	Overweight (BMI 25.1–29.9)	93 (34)
	Obese (BMI $\geq$ 30)	108 (39)
Physical activity n (%)		
	Not active	32 (12)
	Moderately active	146 (53)
	Very active	96 (35)
Diabetes n (%)		46 (17)
Hypertension n (%)		41 (15)
Hs CRP (mg/L) *		3.5 mg (3)
Glucose (mmol/L) *		6.3 (2.5)
Total cholesterol (mmol/L) *		4.9 (0.9)
Urea (mmol/L) *		4.1 (1.5)

* All values fall within normal limits.

3.2. Distribution of four VDR Gene Polymorphisms in Group of 277 from Emirati Population

Figure 1 shows genotypes and allele percentage frequency distribution of four VDR gene polymorphisms in a group of 277 from the Emirati population. The frequencies of VDR genotypes for Bsml are GG = 37%, AG = 44%, and AA = 20.7%.

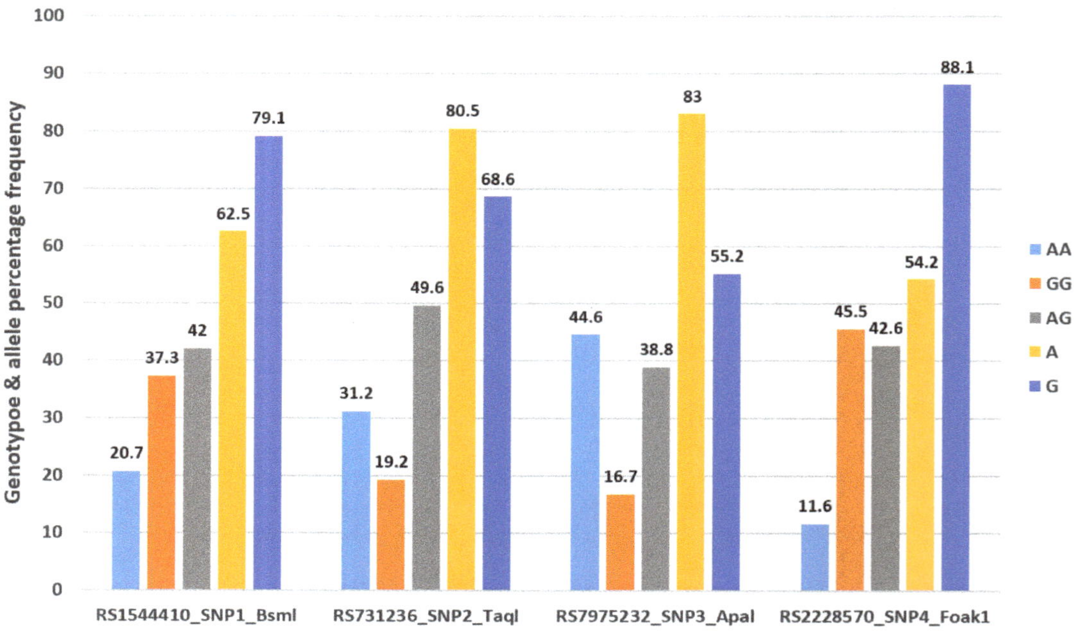

Figure 1. Genotype and allele percentage frequency distribution of four VDR gene polymorphisms in the Emirati population.

3.3. Vitamin D Concentrations According to Distribution and Presence or Absence of Genotypes and Alleles of Four VDR Gene Polymorphisms

Table 2 shows vitamin D concentrations according to genotype distribution of the four VDR gene polymorphisms. There were statistically significant differences in vitamin concentrations between different genotypes of the four VDR gene polymorphisms ($p < 0.05$). Table 3 shows vitamin D concentrations according to the presence or absence of genotypes and allele distribution of the four VDR gene polymorphisms. There was a statistically significant difference in vitamin concentrations between subjects with and without genotypes AA and AG and allele G in Apal VDR gene polymorphisms only ($p < 0.05$).

Table 2. Vitamin D concentrations (ng/mL) according to genotypes distribution of four VDR gene polymorphisms.

	AA	GG	AG	*p* Value
RS1544410_SNP1_Bsml	24.5 (11)	22.1 (11)	23.9 (10)	0.046
RS731236_SNP2_Taql	22.0 (11)	24.0 (11)	23.9 (10)	0.054
RS7975232_SNP3_Apal	25.0 (11)	22.3 (11)	21.7 (9)	0.008
RS2228570_SNP4_Foak1	24.0 (10)	23.9 (12)	22.6 (9)	0.007

Table 3. Mean (SD) vitamin D concentrations (ng/mL) according to presence or absence of genotypes and allele distribution of four VDR gene polymorphisms.

	AA		GG		AG		A		G	
	Yes	No	Yes	No	Yes	No	Yes	No	Yes	No
RS1544410_SNP1_Bsml	25.5 (11)	23.2 (10)	22.1 (11)	24.2 (10)	23.9 (10)	23.1 (11)	23.9 (10)	22.7 (12)	23.0 (10)	25.0 (12)

Table 3. *Cont.*

	AA		GG		AG		A		G	
	Yes	**No**	**Yes**	**No**	**Yes**	**No**	**Yes**	**No**	**Yes**	**No**
RS731236_SNP2_Taql	22.0 (11)	24.1 (10)	24.0 (11)	23.3 (11)	23.9 (10)	23.0 (11)	23.2 (11)	24.5 (11)	23.9 (11)	22.3 (11)
RS7975232_SNP3_Apal	25.1 (11)	22.1 * (10)	22.3 (11)	23.7 (11)	21.7 (9)	24.5 * (11)	23.6 (10)	22.9 (12)	21.9 (10)	25.3 * (12)
RS2228570_SNP4_Foak1	23.9 (10)	23.4 (11)	22.6 (9)	24.1 (11)	23.9 (12)	23.0 (10)	22.9 (10)	24.1 (12)	23.3 (11)	24.6 (11)

* $p < 0.05$ for difference between those with and without the genotypes and allele distribution of four VDR gene polymorphisms.

3.4. Genotype Distribution Based on Gender, BMI and Diabetes, Hypertension

Multiple logistic regression analysis revealed significant and independent association between vitamin D status and age and sex only ($p < 0.05$), (Table 4). We found no statistically significant independent associations between the VDR gene polymorphisms assessed and vitamin D status (deficiency vs. insufficiency/optimal) after adjusting for clinical parameters known to influence vitamin D status including age, sex, dietary intakes, physical activity, sun exposure, smoking and body mass index of study population (Table 4).

Table 4. Multiple logistic regression analysis of the influence of four VDR gene polymorphisms and some clinical prognostic variables on vitamin D status (deficiency vs. insufficiency/optimal) of study population.

	Standardized Coefficients B	**Standard Error**	***p* Value**	**Exp (B)**	**95.0% Confidence Interval for B**	
					Lower Bound	**Upper Bound**
Age (years)	0.018	0.008	0.022	1.018	1.003	1.034
Gender (male/female)	−0.763	0.241	0.002	0.466	0.291	0.748
BMI	−0.006	0.008	0.443	0.994	0.978	1.010
Sun exposure	0.009	0.044	0.838	1.009	0.926	1.099
Diet	0.010	0.036	0.782	1.010	0.941	1.083
Physical activity	0.050	0.154	0.748	1.051	0.777	1.422
Smoking	0.071	0.142	0.616	1.074	0.813	1.417
RS1544410_SNP1_Bsm	0.179	0.146	0.221	1.196	0.898	1.593
RS731236_SNP2_Taql	−0.067	0.127	0.598	0.935	0.730	1.199
RS7975232_SNP3_Apal	−0.079	0.064	0.217	0.924	0.815	1.047
RS2228570_SNP4_	−0.156	0.144	0.279	0.855	0.645	1.135

Figures 2–4 show the genotypes and allele percentage frequency distribution of all four VDR gene polymorphisms, Bsml, Taql, Apal and Fok1 among subjects with and without diabetes, hypertension and obesity. There were no statistically significant differences in the frequency of the genotypes and alleles of Bsml, Taql, Apal and Fok1 VDR genes among patients with diabetes and hypertension compared to those without these medical conditions except in Fok1 AA genotype and G allele (Figures 2 and 3). No statistically significant differences were found in the frequency of the genotypes and alleles of Bsml, Taql, Apal and Fok1 VDR genes in obese and overweight subjects compared to normal weight subjects (Figure 4). Furthermore, no statistically significant difference was found in BMI between subjects with and without the genotype and alleles of all four VDR genes.

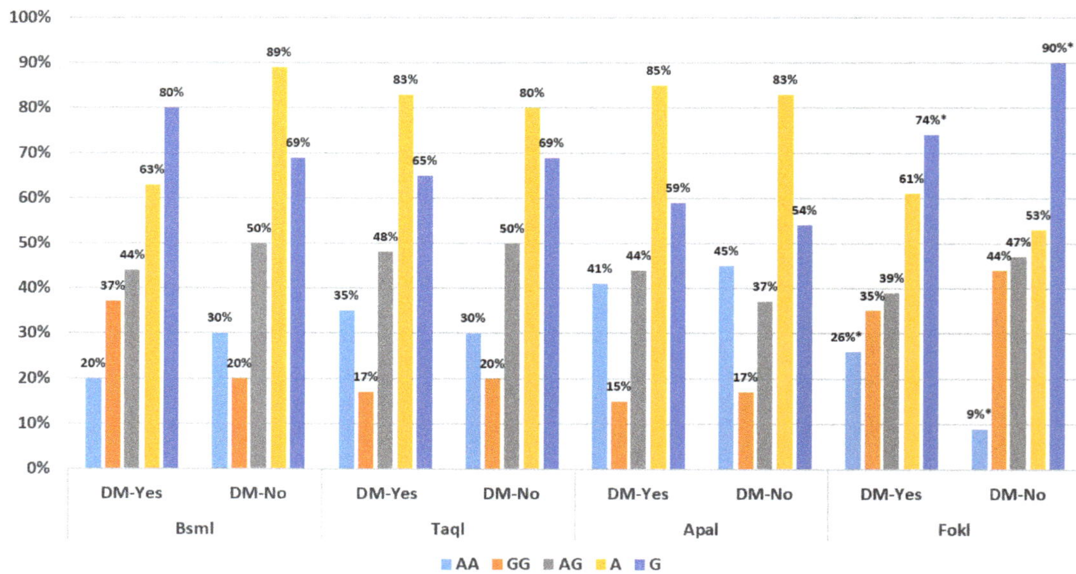

Figure 2. Genotypes and allele percentage distribution of four VDR gene polymorphisms in subject with diabetes (DM-yes) compared with those without diabetes (DM-No). * $p < 0.05$.

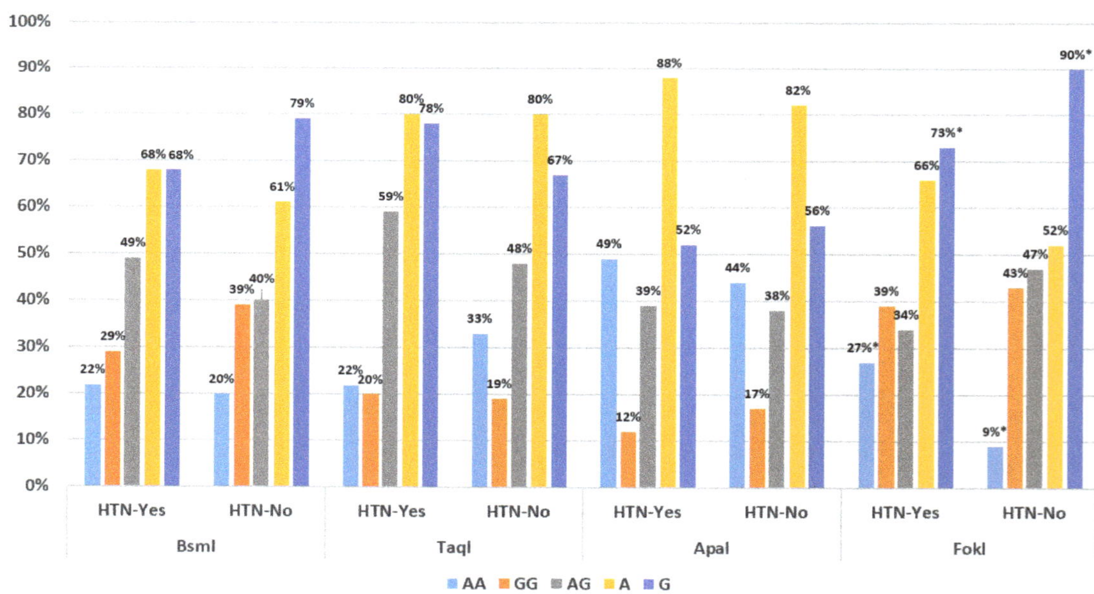

Figure 3. Genotypes and allele percentage distribution of four VDR gene polymorphisms in subject with hypertension (HTN-yes) compared with those without hypertension (HTN-No). * $p < 0.05$.

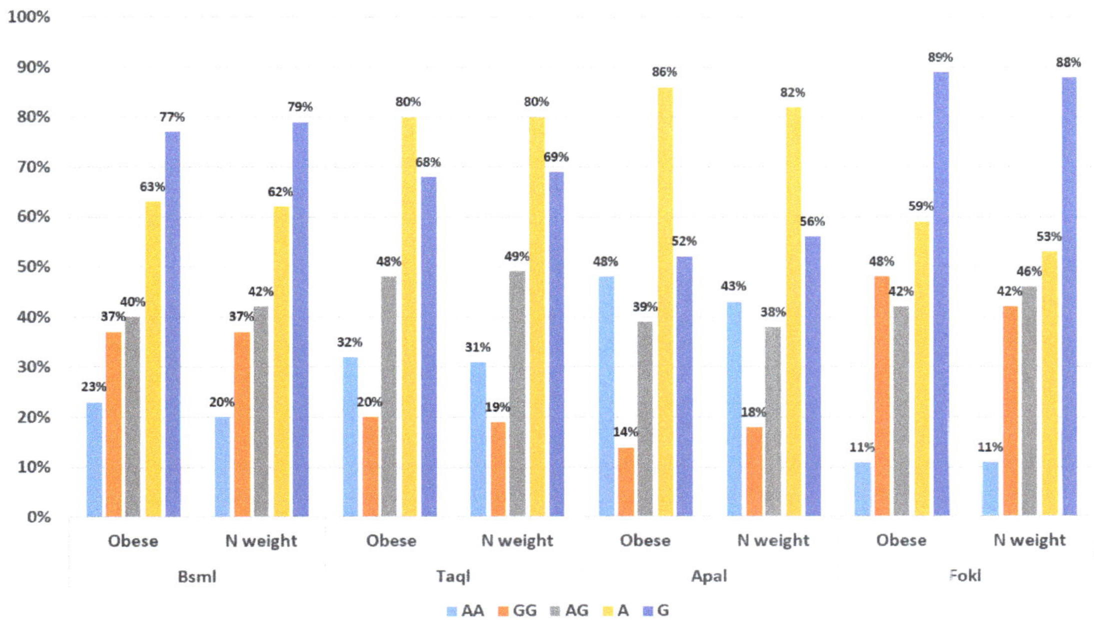

Figure 4. Genotypes and allele percentage distribution of four VDR gene polymorphisms in obese subjects compared to normal weight (N weight) subjects.

4. Discussion

We examined the prevalence of the common four single-nucleotide polymorphisms in the VDR gene including BsmI, FokI, TaqI and ApaI in a large sample of the UAE population and compared vitamin D concentrations between subjects with and without the genotypes and alleles of all four VDR genes. In addition, we compared their association with chronic diseases common in UAE society, including diabetes, hypertension and obesity. Although we found statistically significant differences in vitamin concentrations between different genotypes of the four VDR gene polymorphisms, no association was found after adjusting clinical parameters known to influence vitamin D status. We also found no association with chronic diseases known to be prevalent in UAE society, such as obesity and its associated pathologies including diabetes and hypertensin, except in the Fok1 AA genotype and G allele.

4.1. Genotypes and Allele Percentage Frequency Distribution Compared to Previous Studies

The prevalence of some of the genotypes and alleles is in agreement with previous studies from the UAE community and other societies. For example, the frequency distribution of Bsml is in agreement with a previous study from the UAE (Saadi et al. findings: GG = 37%, AG = 42%, AA = 20%; our findings: GG = 37%, AG = 44%, AA = 20.7%) [17]. For the Taql VDR genotype, our result agree with another study's findings from the UAE and a study from France [15,24]. Although not many studies in the UAE have reported the frequency distribution of the SNP3 Apal, a study from Poland has reported the relationship between SNP3 Apal, vitamin D and the occurrence of cardiovascular disease (CVS) [25]. Our results of SNP4 Fok1 genotypes and allele percentage frequency distribution are different from previous studies, including one from the UAE [15]. A study from India of the frequency of the Fok I and Taq I variants in healthy Indian individuals and its association with 25-OH-Vitamin D levels reported that the distribution of the polymorphic loci Fok I and Taq I vary considerably not only in different populations, but also within India [8].

Overall, the reasons for the difference in vitamin D genotypes and allele nucleotide polymorphisms percentage frequency distribution appear to be differences in sample size and sampling methods used between different studies. The unrepresentative or biased samples made it difficult to make valid generalizations about vitamin D genotypes and allele percentage frequency distribution within and between different populations.

4.2. Vitamin D Concentrations According to the Presence or Absence of Genotypes and Allele Distribution of Four VDR Gene Polymorphism

Although we found statistically significant differences in vitamin concentrations between different genotypes of the four VDR gene polymorphisms, no statistically significant independent associations were found between the four VDR gene polymorphisms assessed and D status after adjusting for clinical parameters known to influence vitamin D status, such as dietary intake, physical activity, sun exposure, smoking and body mass index. Another previous study from the UAE found no association between vitamin D levels and bone turnover markers and BsmI VDR gene polymorphism [17]. In contrast, a study from India reported a significant association between vitamin D levels and Taq 1 SNP but not with the Fok I [8].

4.3. The Prevalence of Vitamin D VDR Receptor Genotypes and Alleles in Subjects with Obesity, Diabetes Mellitus and Hypertension Compared to Those without These Conditions

Circulating vitamin D levels, action and underlying vitamin D receptor VDR genetic polymorphisms have been linked to many common diseases such as obesity and associated pathologies, including diabetes, hypertension and CVS diseases. This is important because of the co-existence of pathologically high proportions of obesity and vitamin D deficiency in our population. Many vitamin D genetic and alleleic polymorphisms have been identified, and their effects on VDR protein function and consequently vitamin D levels and health impact have been studied. The BsmI, FokI, TaqI and ApaI polymorphisms in the VDR gene have been the most studied. In this study we found no statistically significant differences in the percentage frequency distribution of all four VDR gene polymorphisms, Bsml, Taql, Apal and Fok1, among subjects with obesity, diabetes mellitus and hypertension compared to those without these conditions, except in the Fok1 AA genotype and G allele in subjects with diabetes and hypertension compared with those without these conditions. The association between vitamin D receptor VDR polymorphism and the risk of chronic diseases including obesity, hypertension, diabetes and other cardiovascular diseases therefore remains unclear. A recent preliminary study assessed the relationship between the VDR genotypes, plasma concentrations of vitamin D metabolites, and the occurrence of cardiovascular and metabolic disorders in 58 Polish patients treated for various cardiological diseases. Among patients with the TT genotype, frequency of hypertension was higher than among carriers of other ApaI genotypes ($p < 0.01$). In addition, carriers of the TT ApaI, TC TaqI, and GA BsmI genotypes had an increased risk of obesity, while the presence of the FokI TT genotype was associated with a higher incidence of heart failure and hypertension. In conclusion, this study reported that the BsmI AA genotype can be protective against CVS, but this observation needs to be confirmed on a larger group of patients. Particular VDR genotypes were associated with 25-hydroxyvitamin-D levels, and the mechanism of this association should be further investigated [22]. Another cross-sectional study that investigated relationships of vitamin D receptor gene VDR polymorphisms to the components of metabolic syndrome (MetS) including obesity, hypertension, diabetes and dyslipidemia among 198 (148 females, 50 males) Arab adults residing in the United Arab Emirates reported that VDR gene polymorphisms were not associated with MetS. However, the authors reported that it may affect the severity of some of components of MetS, namely the association of BsmI with obesity, FokI and BsmI with dyslipidemia and FokI with systolic blood pressure [26]. Another cross-sectional study from the UAE investigated the association between VDR polymorphisms and type 2 diabetes (T2DM) among 264 patients with T2DM, and 91 healthy controls were enrolled. The TaqI variant was shown to be associated with high cholesterol and LDL-cholesterol levels in T2DM patients, while BsmI was associated with lower BMI and lower LDL cholesterol levels. Their results implied that alleles and haplotypes of the VDR gene are associated with the susceptibility to T2DM in the Emirati population [27]. A meta-analysis of 14 studies of the association of the four polymorphisms FokI, BsmI, ApaI and TaqI in the VDR gene with the susceptibility to T2DM reported no significant associations between BsmI, ApaI and TaqI variants and T2DM risk. The report did suggest an increased risk of T2DM in subjects with the FokI VDR gene polymorphism in an Asian population [28]. The association of low vitamin D level and its receptor polymorphism with obesity was studied in 300 Saudi men. The report suggested that both bb of BsmI and tt of TaqI genotypes were higher in the obese group compared with the lean group and that low vitamin D levels and VDR BsmI and Taq1 genotypes may be a risk factor of obesity [29]. A study from Iran which looked at the association of the VDR gene ApaI, BsmI, and TaqI polymorphisms with obesity in 348 obese and 320 non-obese subjects reported an increased risk of obesity in subjects with VDR ApaI polymorphism. In particular, the A allele and the AA genotype in ApaI were associated with the obesity phenotypes [30]. Another systematic review of the association between genetic polymorphisms and obesity in the Arab world included 59 studies with a total of 15,488 cases and 9760 controls in the final analysis. A total of 76 variants located within or near 49 genes were reported to be significantly associated with obesity. Among the 76 variants, two were described as unique to Arabs, as they have not been previously reported in other populations, and 19 were reported to be distinctively associated with obesity in Arabs but not in non-Arab populations. The authors concluded that there appears to be a unique genetic and clinical susceptibility profile of obesity in Arab patients [31].

4.4. Strengths and Weakness

Our study is one of the very few studies to measure the four single-nucleotide polymorphisms in the VDR gene including BsmI, FokI, TaqI and ApaI in a large sample of the UAE population with a high prevalence of overweight and obesity (76%), diabetes (17%) and hypertension (15%). We used BMI as a measure of adiposity in our study. Although BMI is an easily accessible measure of excessive body weight in the general population, it cannot differentiate between fat and lean body mass. Individuals with normal weight can still demonstrate harmful adiposity traits [32]. Nevertheless, and similar to most previous studies reported on the relationship between vitamin D VDR genetic polymorphisms and risk or severity of chronic diseases, our sample selection and size may not be representative of the whole UAE population. Studies of the relationships between VDR gene polymorphisms and obesity-related diseases need to be powerful enough to detect a statistically significant association if one exists. Most previous studies reported in this area, however, have major heterogeneity and weaknesses in relation to study design, sample size, variables measured and control for confounders. A meta-analysis of these multiple studies may also have a role in clarifying the association. In this study, we also made an attempt to identify a potential independent effect of the four single-nucleotide polymorphisms on vitamin D status by adjusting for a number of clinical parameters likely to affect vitamin D status.

5. Conclusions

Although we found statistically significant differences in vitamin concentrations between different genotypes of the four VDR gene polymorphisms, no independent associations were found between VDR gene polymorphisms assessed and D status (deficiency vs. insufficiency/optimal) after adjusting for clinical parameters known to influence vitamin D status. We also found no association of vitamin D receptor gene polymorphism SNPs with chronic diseases known to be prevalent in UAE society, such as obesity, diabetes and hypertension. There is still, however, the need for large well-designed epidemiological studies to evaluate the relationships between VDR gene polymorphisms and obesity-related diseases in an appropriate representative sample of the target population to which the findings will be referred. This is because of the co-existence of pathologically high proportions of obesity and vitamin D deficiency conditions in our society and the Middle East at large.

Author Contributions: S.G. wrote the first draft, S.G. and J.Y. contributed to the design, running of study, analysis and presentation of data. G.S.M.A.-B. contributed data management and analysis. All authors have read and agreed to the published version of the manuscript.

Funding: This study was supported by a grant from United Arab Emirates University project Grant (NP-17-11). The funder had no role on the study design, data analysis, interpretation and writing of the manuscript.

Institutional Review Board Statement: Al Ain Medical District Human research ethics committee approved the study protocol and consent to participate form (Reference number: AAMDHREC protocol no 14/15).

Informed Consent Statement: An informed written consent was obtained from all patients recruited to this study.

Data Availability Statement: Data is available upon request to the corresponding author.

Conflicts of Interest: The authors declare that they have no financial or non-financial competing interest or conflict of interest.

Abbreviations

BMI = body mass index
CVS = cardiovascular disease
ECL = chemiluminescence
LDL = Low density lipoprotein
SNPs = single nucleotide polymorphism
T2DM = type 2 diabetes mellitus
UAE = United Arab Emirates
VDR = vitamin D receptor
MetS = metabolic syndrome

References

1. Holick, M.F. Vitamin D deficiency. *N. Engl. J. Med.* **2007**, *357*, 266–281. [CrossRef]
2. Uitterlinden, A.G.; Fang, Y.; Van Meurs, J.B.; Pols, H.A.; Van Leeuwen, J.P. Genetics and biology of vitamin D receptor polymorphisms. *Gene* **2004**, *338*, 143–156. [CrossRef] [PubMed]
3. Rezende, V.B.; Barbosa, F.; Montenegro, M.F.; Sandrim, V.C.; Gerlach, R.F.; Tanus-Santos, J.E. An interethnic comparison of the distribution of vitamin D receptor genotypes and haplotypes. *Clin. Chim. Acta* **2007**, *384*, 155–159. [CrossRef]
4. Qin, W.H.; Wang, H.X.; Qiu, J.L.; Huang, X.B.; Huang, Y.; Wu, N.R.; Liang, H.S. A meta-analysis of association of vitamin D receptor BsmI gene polymorphism with the risk of type 1 diabetes mellitus. *J. Recept. Signal Transduct. Res.* **2014**, *34*, 372–377. [CrossRef] [PubMed]
5. Al-Daghri, N.M.; Al-Attas, O.S.; Alkharfy, K.M.; Khan, N.; Mohammed, A.K.; Vinodson, B.; Ansari, M.G.A.; Alenad, A.; Alokail, M.S. Association of VDR-gene variants with factors related to the metabolic syndrome, type 2 diabetes and vitamin D deficiency. *Gene* **2014**, *542*, 129–133. [CrossRef]
6. Mosaad, Y.M.; Hammad, E.M.; Fawzy, Z.; Aal, I.A.A.; Youssef, H.M.; ElSaid, T.O.; Monir, R.; El-Deek, B.S. Vitamin D receptor gene polymorphism as possible risk factor in rheumatoid arthritis and rheumatoid related osteoporosis. *Hum. Immunol.* **2014**, *75*, 452–461. [CrossRef]
7. Neela, V.S.K.; Suryadevara, N.C.; Shinde, V.G.; Pydi, S.S.; Jain, S.; Jonnalagada, S.; Singh, S.S.; Valluri, V.L.; Anandaraj, M. Association of Taq I, Fok I and Apa I polymorphisms in Vitamin D Receptor (VDR) gene with leprosy. *Hum. Immunol.* **2015**, *76*, 402–405. [CrossRef]
8. Bhanushali, A.A.; Lajpal, N.; Kulkarni, S.S.; Chavan, S.S.; Bagadi, S.S.; Das, B.R. Frequency of fokI and taqI polymorphism of vitamin D receptor gene in Indian population and its association with 25-hydroxyvitamin D levels. *Indian J. Hum. Genet.* **2009**, *15*, 108–113. [CrossRef]
9. McKinsey Global Institute. Overcoming Obesity: An Initial Economic Analysis. 2021. Available online: http://www.mckinsey.com/insights/economic_studies/how_the_world_could_better_fight_obesity (accessed on 1 March 2023).
10. Fu, J.; Han, L.; Zhao, Y.; Li, G.; Zhu, Y.; Li, Y.; Li, M.; Gao, S.; Willi, S.M. 25(OH) D levels are associated with metabolic syndrome in adolescents and young adults: The BCAMS study. *Clin. Nutr.* **2019**, *38*, 2161–2167. [CrossRef] [PubMed]

11. Palaniswamy, S.; Gill, D.; De Silva, N.M.; Lowry, E.; Jokelainen, J.; Karhu, T.; Mutt, S.J.; Dehghan, A.; Sliz, E.; Chasman, D.I.; et al. Could vitamin D reduce obesity-associated inflammation? Observational and Mendelian randomization study. *Am. J. Clin. Nutr.* **2020**, *111*, 1036–1047. [CrossRef]
12. Cade, C.; Norman, A.W. Vitamin D3 improves impaired glucose tolerance and insulin secretion in the vitamin D-deficient rat in vivo. *Endocrinology* **1986**, *119*, 84–90. [CrossRef]
13. Earthman, C.P.; Beckman, L.M.; Masodkar, K.; Sibley, S.D. The link between obesity and low circulating 25-hydroxyvitamin D concentrations: Considerations and implications. *Int. J. Obes.* **2012**, *36*, 387–396. [CrossRef] [PubMed]
14. Yang, L.; Wu, L.; Fan, Y.; Ma, J. Associations among Four Polymorphisms (BsmI, FokI, TaqI and ApaI) of Vitamin D Receptor Gene and End-stage Renal Disease: A Meta-Analysis. *Arch. Med. Res.* **2015**, *46*, 1–7. [CrossRef] [PubMed]
15. Osman, E.; Al Anouti, F.; El Ghazali, G.; Haq, A.; Mirgani, R.; Al Safar, H. Frequency of rs731236 (Taql), rs2228570 (Fok1) of Vitamin-D Receptor (VDR) gene in Emirati healthy population. *Meta Gene* **2015**, *6*, 49–52. [CrossRef] [PubMed]
16. Gupta, A. Vitamin D deficiency in India: Prevalence, causalities and interventions. *Nutrients* **2014**, *6*, 729–775. [CrossRef]
17. Saadi, H.F.; Nagelkerke, N.; Benedict, S.; Qazaq, H.S.; Zilahi, E.; Mohamadiyeh, M.K.; Al-Suhaili, A.I. Predictors and relationship of serum 25 hydroxyvitamin D concentration with bone turnover markers, bone mineral density and vitamin D receptor genotype in Emirati women. *Bone* **2006**, *39*, 1136–1143. [CrossRef] [PubMed]
18. Holvik, K.; Meyer, H.E.; Søgaard, A.J.; Selmer, R.; Haug, E.; Falch, J.A. Biochemical markers of bone turnover and their relation to forearm bone mineral density in persons of Pakistani and Norwegian background living in Oslo, Norway: The Oslo Health Study. *Eur. J. Endocrinol.* **2006**, *155*, 693–699. [CrossRef]
19. Lowe, N.M.; Mitra, S.R.; Foster, P.C.; Bhojani, I.; McCann, J.F. Vitamin D status and markers of bone turnover in Caucasian and Southy Asian postmenopausal women living in the UK. *Br. J. Nutr.* **2010**, *103*, 1706–1710. [CrossRef]
20. Farrar, M.D.; Kift, R.; Felton, S.J.; Berry, J.L.; Durkin, M.T.; Allan, D.; Vail, A.; Webb, A.R.; Rhodes, L.E. Recommended summer sunlight exposure amounts fail to produce sufficient vitamin D status in UK adults of South Asian origin. *Am. J Clin. Nutr.* **2011**, *94*, 1219–1224. [CrossRef] [PubMed]
21. Saadi, H.F.; Dawodu, A.; Afandi, B.O.; Zayed, R.; Benedict, S.; Nagelkerke, N. Efficacy of daily and monthly high-dose calciferol in vitamin D–deficient nulliparous and lactating women. *Am. J. Clin. Nutr.* **2007**, *85*, 1565–1571. [CrossRef]
22. Laaksonen, M.; Kärkkäinen, M.; Outila, T.; Vanninen, T.; Ray, C.; Lamberg-Allardt, C. Vitamin D receptor gene BsmI-polymorphism in Finnish premenopausal and postmenopausal women: Its association with bone mineral density, markers of bone turnover, and intestinal calcium absorption, with adjustment for lifestyle factors. *J. Bone Miner. Metab.* **2002**, *20*, 383–390. [CrossRef] [PubMed]
23. Gariballa, S.; Shah, I.; Yasin, J.; Alessa, A. Vitamin D [25f(OH)D] metabolites and epimers in obese subject: Interaction and correlations with adverse metabolic health risk factors. *J. Steroid Biochem. Mol. Biol.* **2022**, *215*, 106023. [CrossRef] [PubMed]
24. Zmuda, J.M.; Cauley, J.A.; Danielson, M.E.; Wolf, R.L.; Ferrell, R.E. Vitamin D receptor gene polymorphisms, bone turnover, and rates of bone loss in older African-American women. *J. Bone Miner. Res.* **1997**, *12*, 1446–1452. [CrossRef]
25. Abouzid, M.; Kruszyna, M.; Burchardt, P.; Kruszyna, Ł.; Główka, F.K.; Karaźniewicz-Łada, M. Vitamin D Receptor Gene Polymorphism and Vitamin D Status in Population of Patients with Cardiovascular Disease-A Preliminary Study. *Nutrients* **2021**, *13*, 3117. [CrossRef]
26. Hasan, H.A.; Ra'ed, O.A.; Muda, W.A.M.B.W.; Mohamed, H.J.B.J.; Samsudin, A.R. Association of Vitamin D receptor gene polymorphisms with metabolic syndrome and its components among adult Arabs from the United Arab Emirates. *Diabetes Metab. Syndr.* **2017**, *11* (Suppl. S2), S531–S537. [CrossRef]
27. Al Safar, H.; Chehadeh, S.E.H.; Abdel-Wareth, L.; Haq, A.; Jelinek, H.F.; ElGhazali, G.; Al Anouti, F. Vitamin D receptor gene polymorphisms among Emirati patients with type 2 diabetes mellitus. *J. Steroid Biochem. Mol. Biol.* **2018**, *175*, 119–124. [CrossRef]
28. Li, L.; Wu, B.; Liu, J.Y.; Yang, L.B. Vitamin D receptor gene polymorphisms and type 2 diabetes: A meta-analysis. *Arch. Med. Res.* **2013**, *44*, 235–241. [CrossRef]
29. Al-Hazmi, A.S.; Al-Mehmadi, M.M.; Al-Bogami, S.M.; Shami, A.A.; Al-Askary, A.A.; Alomery, A.M.; Al-Shehri, S.S.; Dahlawi, H.; Abdulrazag, K.; Ali, T.; et al. Vitamin D receptor gene polymorphisms as a risk factor for obesity in Saudi men. *Electron. Physician* **2017**, *9*, 5427–5433. [CrossRef] [PubMed]
30. Rashidi, F.; Ostadsharif, M. Association of VDR gene Apal polymorphism with obesity in Iranian population. *Biomedica* **2021**, *41*, 651–659. [CrossRef]
31. Younes, S.; Ibrahim, A.; Al-Jurf, R.; Zayed, H. Genetic polymorphisms associated with obesity in the Arab world: A systematic review. *Int. J. Obes.* **2021**, *45*, 1899–1913. [CrossRef]
32. Sahakyan, K.R.; Somers, V.K.; Rodriguez-Escudero, J.P.; Hodge, D.O.; Carter, R.E.; Sochor, O.; Coutinho, T.; Jensen, M.D.; Roger, V.L.; Singh, P.; et al. Normal-Weight Central Obesity: Implications for Total and Cardiovascular Mortality. *Ann. Intern. Med.* **2015**, *163*, 827–835. [CrossRef] [PubMed]

biomedicines

MDPI

Article

Effect of Single High Dose Vitamin D Substitution in Hospitalized COVID-19 Patients with Vitamin D Deficiency on Length of Hospital Stay

Fabienne Jaun [1,2,3,*], Maria Boesing [1,2], Giorgia Luethi-Corridori [1,2], Kristin Abig [1,2], Nando Bloch [4], Stéphanie Giezendanner [1,2], Victoria Grillmayr [4], Philippe Haas [1], Anne B. Leuppi-Taegtmeyer [1,2], Jürgen Muser [5], Andrea Raess [4], Philipp Schuetz [2,6], Michael Brändle [4] and Jörg D. Leuppi [1,2,*]

1 University Center of Internal Medicine, Cantonal Hospital Baselland, 4410 Liestal, Switzerland; maria.boesing@ksbl.ch (M.B.); giorga.luethi-corridori@ksbl.ch (G.L.-C.); kristin.abig@ksbl.ch (K.A.); philippehaas@mac.com (P.H.); anne.leuppi-taegtmeyer@usb.ch (A.B.L.-T.)
2 Medical Faculty, University of Basel, 4001 Basel, Switzerland; philipp.schuetz@ksa.ch
3 Paracelsus Medical University, Master Programme Public Health, Center for Public Health and Healthcare Research, 5020 Salzburg, Austria
4 Cantonal Hospital St. Gallen, 9000 St. Gallen, Switzerland; nando.bloch@kssg.ch (N.B.); victoria.grillmayr@kssg.ch (V.G.); andrea.raess@kssg.ch (A.R.); michael.brandle@kssg.ch (M.B.)
5 Central Laboratories, Cantonal Hospital Baselland, 4410 Liestal, Switzerland
6 Cantonal Hospital Aarau, 5001 Aarau, Switzerland
* Correspondence: fabienne.jaun@ksbl.ch (F.J.); joerg.leuppi@ksbl.ch (J.D.L.)

Abstract: Vitamin D and its role in the coronavirus-19 disease (COVID-19) pandemic has been controversially discussed, with inconclusive evidence about vitamin D3 (cholecalciferol) supplementation in COVID-19 patients. Vitamin D metabolites play an important role in the initiation of the immune response and can be an easily modifiable risk factor in 25-hydroxyvitamin D_3 (25(OH)D_3)-deficient patients. This is a multicenter, randomized, placebo-controlled double-blind trial to compare the effect of a single high dose of vitamin D_3 followed by treatment as usual (TAU) of daily vitamin D_3 daily until discharge versus placebo plus TAU in hospitalized patients with COVID-19 and 25(OH)D_3-deficiency on length hospital stay. We included 40 patients per group and did not observe a significant difference in the median length of hospital stay (6 days in both groups, $p = 0.920$). We adjusted the length of stay for COVID-19 risk factors ($\beta = 0.44$; 95% CI: −2.17–2.22), and center ($\beta = 0.74$; 95% CI: −1.25–2.73). The subgroup analysis in patients with severe 25(OH)D_3-deficiency (<25 nmol/L) showed a non-significant reduction in the median length of hospital stay in the intervention group (5.5 vs. 9 days, $p = 0.299$). The competing risk model with death did not reveal significant differences between the group in the length of stay (HR = 0.96, 95% CI 0.62–1.48, $p = 0.850$). Serum 25(OH)D_3 level increased significantly in the intervention group (mean change in nmol/L; intervention: +26.35 vs. control: −2.73, $p < 0.001$). The intervention with 140,000 IU vitamin D_3 + TAU did not significantly shorten the length of hospital stay but was effective and safe for the elevation of serum 25(OH)D_3 levels.

Keywords: vitamin D; COVID-19; SARS-CoV-2; vitamin D deficiency; vitamin D substitution; immunomodulation

1. Introduction

For more than two years, the world has been facing a global pandemic, with a then new coronavirus (CoV). The underlying pathogen of this new disease was later identified as Severe Acute Respiratory Syndrome Coronavirus-2 (SARS-CoV-2) [1]. Within a short time period, the virus spread quickly around the world, and led the World Health Organization (WHO) to reclassify the epidemic as a global pandemic in March 2020 [2].

The symptoms that occur in a COVID-19 infection are similar to those shown in patients with SARS or MERS, which is explained by the genetic relationship of the virus [3]. Risk factors for a poor outcome of COVID-19 have been found to include older age, male sex, obesity, chronic respiratory conditions, cardiovascular disease (CVD), metabolic diseases, malignancies, and active smoking [4–8].

There are studies that support the theory that 25(OH)D_3 deficiency plays a role in the progression and severity of COVID-19, other studies could not find a significant connection between the two conditions [9]. However, the scientific community has not yet found a consensus about the role of vitamin D metabolites in the current pandemic. This randomized, placebo-controlled, double-blind trial investigated if a single high dose of vitamin D_3, in addition to standard dosed vitamin D_3 supplementation, can positively influence the length of hospital stay in patients with 25(OH)D_3 deficiency and COVID-19 compared to patients receiving the treatment as usual for 25(OH)D_3 deficiency only.

Deficiency of 25(OH)D_3 is a well-documented phenomenon that occurs in approximately 40% to 60% of the population, depending on whether the threshold is 50 nmol/L or 75 nmol/L as defining criteria for deficiency [10]. There are observational studies reporting independent associations between low serum concentrations of 25(OH)D_3 and susceptibility to acute respiratory tract infection [11,12]. 25(OH)D_3 supports the induction of antimicrobial peptides in response to both viral and bacterial stimuli, suggesting a potential mechanism by which vitamin D inducible protection against respiratory pathogens might be mediated [13–15].

Various studies have been conducted that investigated the influence and role of vitamin D_3 supplementation in various dosages (from medium dose to ultra-high dose) in patients with critical illnesses. A retrospective observational study of the serum 25(OH)D_3 status in patients with ARDS, which is a common complication in patients with COVID-19, showed that over 95% of these patients had 25(OH)D_3 deficiency (n = 108) [16]. So far, the influence of vitamin D_3 in the pandemic is unclear, the current level of evidence is low and contradictory. This might be influenced by various factors, such as the lack of high-quality evidence, different methodological approaches, or inconsistent enrolment criteria. Some studies found significant differences on various variables, such as the time to resolve symptoms or biomarker, mortality, or ICU admission [17–21], whereas others did not find any relationship between those outcomes and vitamin D_3 supplementation [22–24]. Most of the studies or reviews concluded that further research in the form of randomized controlled trials is indicated to clarify the impact of vitamin D_3 supplementation in patients with COVID-19, as there are still unanswered questions and uncertainties that need clarification. Previous findings in vitamin D research found a reduction in ventilation days for critically ill patients, a better muscle tissue function, and suppressed cytokines in patients who were treated with high dose vitamin D_3 [25]. According to previous studies, there was no increased risk for adverse events or side effects observed when patients were treated with high dose vitamin D [26,27]. Further, there are observations that a single high dose of vitamin D is more effective in elevating the serum 25(OH)D_3 levels, whereas daily administration is more beneficial for health outcomes [28]. We therefore came to hypothesize that a single high dose of vitamin D_3 in addition to standard treatment improves the recovery period positively in patients with COVID-19 and 25(OH)D_3 deficiency compared to daily smaller doses of vitamin D_3 only. That means that the time to recovery, which was defined as length of hospital stay, is shorter in the single high dose vitamin D_3 group relative to standard treatment group only.

2. Materials and Methods

This is a multicenter, randomized, placebo-controlled double-blind trial to compare the effect of a single high dose of vitamin D_3 (140,000 IU, Vitamin D3 Wild Oel 500 IU/drop, Muttenz, Switzerland) followed by treatment as usual (TAU) of daily vitamin D_3 (800 IU, Vitamin D3 Streuli 4000 IU/ml, Streuli Pharma AG, Uznach, Switzerland) until discharge versus placebo plus TAU in hospitalized patients with COVID-19 and 25(OH)D_3 deficiency on length hospital stay. All patients that were hospitalized due to COVID-19 in one of the participating centers (Cantonal hospital Baselland and Cantonal hospital St. Gallen) were considered as potentially eligible for participation in this trial and therefore checked for the inclusion and exclusion criteria (Table 1). The patients were recruited between December 2020 and August 2021. Participants were randomized in a 1:1 ratio to either group and randomization was stratified by center using block randomization (block size = 4).

Table 1. Inclusion and exclusion criteria.

Inclusion Criteria
Informed consent as documented by signature
>18 years old
Ongoing COVID-19 infection (PCR confirmed)
Hospitalized in one of the participating study centers
Laboratory-confirmed serum 25(OH)D_3 < 50 nmol/L(<20 ng/mL)
Exclusion criteria
Known hypersensitivity against one of the used vitamin D_3 products
Active malignancy
Hypercalcemia defined as serum calcium >2.2 mmol/L
Granulomatous disease (e.g., sarcoidosis)
History of renal stones in the past 12 months
Pregnancy or breastfeeding
Admission to ICU before inclusion or at day of inclusion

To elevate the serum 25(OH)D_3 level quickly and sufficiently, a single dose of 140,000 IU (3500 µg) vitamin D_3 was administered orally. This dose was found to be safe and effective in a study of high-dose vitamin D_3 in patients with tuberculosis [29,30]. The dosage recommended for the standard treatment of a vitamin D_3 deficiency varies from 400 to 1000 IU per day [31]. In consultation with a pharmacovigilance expert (University Hospital of Basel), the study team considered that the dose of 800 IU was adequate to elevate the serum 25(OH)D_3 level in the included patients.

2.1. Outcomes

The primary outcome was defined as the length of the hospital stay (LoS), defined as the time from randomization to discharge. Further, the time from symptom onset to discharge and the time from hospital admission to discharge were assessed and analyzed.

As secondary outcomes, we defined COVID-19-related complications, namely ICU admission, mechanical ventilation, death, and other complications linked to COVID-19. Additionally, we looked at the serum 25(OH)D_3 levels, the health-related quality of life on days 28 and 90, self-reported symptoms, and vital signs. For safety outcomes, the occurrence of adverse events (AEs) and serious adverse events (SAEs) were assessed [32].

2.2. Data Collection and Management

Each participant underwent the same procedure regarding the study-related visits from inclusion until discharge and for the follow-up period (Supplementary Table S1). The patients had daily visits until discharge by treating physicians and nurses. The Short Form 12 item health survey (SF-12) to assess the health=related quality of life (hrQoL) on day 28 and day 90 days after randomization was assessed by a telephone visit. The data were collected in an electronic data capture system (SecuTrial®), which is accessible via a standard browser and personalized and password protected.

2.3. Statistical Analysis

To perform the statistical analysis, either R Statistical software (Version 4.0.3) or SPSS Statistics, (Version 24 (IBM)) was used. We performed a sample size calculation based on the assumption of a true difference in the mean length of stay of two days (SD = 2.96) [33].

According to the calculation, accounting for 10% dropouts and the chosen block size randomization, a total number of 80 patients (40 per group) were included.

For the primary outcome, we used the Mann–Whitney-U, due to non-normal distribution, unless our assumptions for the sample size calculations were based on a two-sided *t*-test. For continuous data, independent two-sample Student *t*-test or Mann–Whitney-U test was used for secondary analyses of group differences in the variables of interest. To compare the proportion of patients between the two groups regarding the secondary outcomes, a Chi^2 test was performed and if values had expected frequencies <5, the Fisher exact test was used with simulated p-value (based on 2000 replicates). To evaluate the change in the SF-12 questionnaire over three time points, linear mixed effects models were used to determine time-by-treatment interactions using the Satterthwaite approximation to estimate *p*-values. Sensitivity analysis of the primary outcome was performed with a robust linear regression model adjusted for known COVID-19 risk factors. To evaluate the safety of short-term high dose vitamin D_3 administration in the above-described population, linear/logistic mixed-effect models were used to estimate treatment effects on adverse and serious adverse events across time.

A competing risk analysis that accounts for death as competing risk (besides the event of interest which was time to discharge) using the Fine and Gray model and a subgroup analysis for patients who had severe vitamin D deficiency at randomization, defined as 25(OH)D_3 level <25 nmol/L, for the primary endpoint with the same method used in the full dataset was performed post hoc. We only performed the binary analysis for the subgroup based on the small number of patients in the subgroup. The detailed methodology of the trial has been published previously [34].

3. Results

A total of 80 patients were randomized, 40 patients in each group. Two patients were excluded after randomization because the serum 25(OH)D_3 level at the day of inclusion was >50 nmol/L. The data of 78 patients were used for the analysis. A total of 39 patients were in the intervention group and 39 patients in the placebo group. Median time from hospital admission to randomization was two days for both groups (SD ± 1 day).

3.1. Clinical Characteristics at Time of Randomization

The baseline characteristics of our population are summarized in Table 2. The mean age of our population was 60.5 (range 35–91) years in the intervention group and 61.5 (range 29–93) years in the control group. There were more women in the intervention group (35.9%) compared to the control group (17.9%; $p = 0.125$). Mean 25(OH)D_3 level at time of randomization was 31.03 (SD ± 10.95) nmol/L in the intervention group and 28.54 (±10.13) nmol/L for the control group ($p = 0.232$). The two groups did not differ in terms of mean vital signs on the day of randomization. The mean duration from symptom onset to hospital admission was 8 days (range 3–18) in the intervention group versus 7 days (range 1–19) in the control group ($p = 0.335$). Almost all patients (intervention: 100% vs. control: 95%) showed COVID-19 typical changes in the radiological imaging of the lungs ($p = 0.152$). The intervention group had a significantly higher mean CRP value at the time of randomization compared to the control group (intervention: 96.50 mg/L vs. control: 66.67 mg/L, $p = 0.015$); no other value in the hematology profile or blood chemistry showed a significant difference between the groups. The mean SF-12 mental score (hrQoL sub score) was significantly higher in the control group compared to the intervention group (intervention: 41.76 ± 15.51 versus control: 51.66 ± 8.04; $p = 0.013$). In terms of co-morbidities at the time of randomization and concomitant medication during the hospitalization, no significant differences between the groups were observed (Supplementary Table S2, Supplementary Table S3).

Table 2. Demographic and baseline characteristics of the included patients at time of randomization.

Baseline Characteristics		N [a] (Intervention/Control)	Intervention	Control	p-Value
Age	years	39/39	60.49 ± 13.84	61.38 ± 15.29	0.787
Sex	female, % (n)	39/39	35.9 (14)	17.9 (7)	0.125 ‡
BMI	kg/m^2	39/38	28.55 ± 5.09	28.44 ± 5.68	0.951 †
RR	/min	33/34	24 ± 5	23 ± 6	0.362 †
SpO_2	%	37/37	92 ± 2.5	93 ± 2.5	0.119
Patient with dyspnea [b]	%	6/13	92 ± 2.5	92 ± 3	0.487
Oxygen supplementation requirement	% (n)	38/39	37.8 (14)	62.2 (23)	0.084
Patients with dyspnea [b]	% (n)	6/13	50 (3)	71.4 (10)	0.393
O_2 -supplementation	L/min	36/35	1.5 (±2)	2 (±2)	0.071 †
Patients with dyspnea [b]	L/min	6/13	2 ± 2.5	3 ± 3	0.610
Symptom onset to hospitalization	days	39/39	8 ± 4	7 ± 4	0.335
Radiological evidence	% (n)	39/39	100 (39)	94.9 (37)	0.152
Quality of life [c]		26/20			
SF-12 physical score			31.39 ± 10.61	27.85 ± 11.75	0.352 †
SF-12 mental score			41.76 ± 15.51	51.66 ± 8.04	0.018 †
Hematology					
Hemoglobin	g/L	30/34	129.67 ± 15.89	134.68 ± 15.78	0.211
RBC	$10^6/\mu L$	31/34	4.35 ± 0.43	4.353 ± 0.74	0.440 †
Leukocytes	$10^3/\mu L$	31/34	7.11 ± 2.46	7.647 ± 3.23	0.451
Thrombocytes	$10^3/\mu L$	30/34	244.90 ± 81.02	238.56 ± 101.88	0.783
Blood chemistry					
25(OH)D3	nmol/L	37/39	31.46 ± 10.95	28.54 ± 10.13	0.232
PTH	pmol/L	7/6	3.14 ± 1.34	4.160 ± 3.93	0.943 †
Calcium	mmol/L	9/13	2.15 ± 0.06	2.09 ± 0.34	0.282 †
Phosphate	mmol/L	7/12	1.00 ± 0.29	0.86 ± 0.27	0.570
CRP [d]	mg/L	28/31	96.50 ± 57.07	66.67 ± 61.23	0.015 †

Abbreviations: ALP = alkaline phosphatase, ALT = alanine aminotransferase, AST = aspartate aminotransferase, BMI = body mass index, BP = blood pressure, CRP = C-reactive protein, eGFR = estimated glomular filtration rate, MCV = mean corpuscular volume, SpO_2 = peripheral oxygen saturation, RBC = red blood cells, RR = respiratory rate. Continuous variables: mean ± standard deviation (SD). Categorical variables: percentages of patients (absolute number of patients per group = n). [a] number of patients with available information. [b] Values for patients with self-reported dyspnea. [c] Assessment of the quality of life started when the trial was already ongoing, therefore not all patients have a baseline value. [d] Mean values are outside the normal range. ‡ Fisher's exact test when expected frequencies <5 instead of Chi^2 test for categorical variables. † Mann–Whitney-U-Test.

3.2. Primary Outcome

There was no significant difference between intervention group and control group in terms of length of hospital stay (Table 3, Figure 1). Median time from symptom onset to discharge was 16 days for the intervention group and 15 days for the control group ($p = 0.193$). Median time from hospital admission to discharge was 8 days for both groups ($p = 0.924$) and the median time from randomization to discharge was 6 days for both groups ($p = 0.920$).

We adjusted the length of stay variables for COVID-19 risk factors, including age, number of comorbidities, peripheral oxygen saturation, number of symptoms, and 25(OH)D_3 value at randomization ($\beta = 0.03$, 95% CI: −2.17–2.22), prognostic imbalances between the groups (adjusted for $p < 0.010$) ($\beta = 0.39$, 95% CI: −1.78–2.56), and center ($\beta = 0.74$, 95% CI: −1.25–2.73) (Supplementary Table S4 Appendix B).

Table 3. Length of hospital stay per group.

Length of Stay in Days	Intervention, n = 39 Median (IQR)	Control, n = 39 Median (IQR)	*p*-Value
Time from symptom onset to discharge	16 (9)	15 (8)	0.193 †
Time from hospital admission to discharge	8 (8)	8 (6)	0.924 †
Time from randomization to discharge	6 (10)	6 (6)	0.920 †

† Mann–Whitney-U-Test.

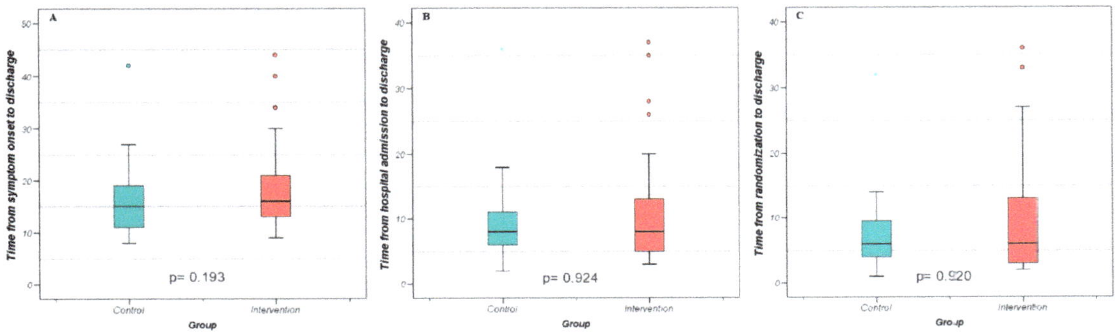

Figure 1. Length of hospital stay as boxplots over the variables "time from symptom onset to discharge", "time from hospital admission to discharge", and "time from randomization to discharge" with n = 78 (39 patients per group). (**A**). "Time from symptom onset to discharge" in days (intervention in blue and control in red), $p = 0.193$. The dots are outliers (1.5 IQR). (**B**). "Time from hospital admission to discharge" (intervention in blue and control in red), $p = 0.924$. The dots indicating the outliers (1.5 IQR). The blue star is an extreme value (3 IQR). (**C**). "Time from randomization to discharge" (intervention in blue and control in red), $p = 0.920$. The dots are outliers (1.5 IQR). The blue star is an extreme value (3 IQR).

Post Hoc Analysis

The post-hoc analysis of patients with severe 25(OH)D_3 deficiency showed a median time from symptom onset to discharge of 15 days in both groups ($p = 0.764$). Median time from hospital admission to discharge was 5.5 days for the intervention group and 9 days for the control group ($p = 0.299$) and the median time from randomization to discharge was 4.5 days for the intervention group and 7 days for the control group ($p = 0.444$) (Table 4).

Table 4. Length of hospital stay in patients with severe vitamin D deficiency.

Length of Stay (Days) When Serum 25(OH) D Level <25 nmol /L, n = 25	Intervention, n = 12 Median (IQR)	Control, n = 13 Median (IQR)	*p*-Value
Time from symptom onset to discharge	15 (11)	15 (9)	0.764 †
Time from hospital admission to discharge	5.5 (13)	9 (6)	0.299 †
Time from randomization to discharge	4.5 (13)	7 (5)	0.444 †

† Mann–Whitney-U-Test.

In the competing risks model with the full dataset and death as the competing risk, groups did not differ in (a) time from symptom onset to discharge (HR = 0.88, 95% CI 0.57–1.36, p = 0.570), (b) time from hospital admission to discharge (HR = 0.99, 95% CI 0.64–1.52, p = 0.950), or (c) time from randomization to discharge (HR = 0.96, 95% CI 0.62–1.48, p = 0.850) Figure 2.

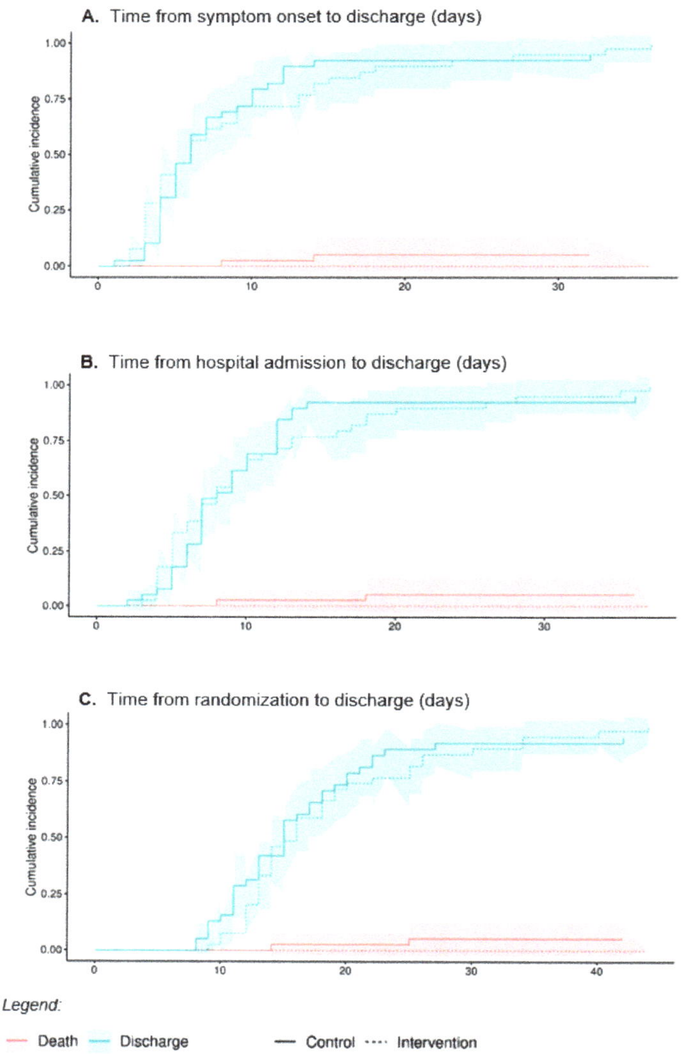

Figure 2. Competing risk model length of stay with death as competing risk. The blue lines indicate the event "discharge", whereas the red indicates the competing risk "death", with its 95% confidence interval. (**A**). Time from symptom onset to discharge, event discharge: HR 0.88 (95% CI 0.57–1.36), p = 0.570. Event Death: HR 0.47 (95% CI: 0.4–5.02), p = 0.540. (**B**). Time from hospital admission to discharge, event discharge: HR 0.99 (95% CI 0.64–1.52), p = 0.950. Event Death: HR 0.49 (95% CI: 0.05–5.16), p = 0.550. (**C**). Time from randomization to discharge, event discharge: HR 0.69 (95% CI 0.62–1.48), p = 0.850. Event Death: HR 0.49 (95% CI: 0.05–5.16), p = 0.550.

3.3. Secondary Outcomes

In each group, four patients were admitted to ICU (intervention: 10.3% (n = 4) vs. control: 10.3% (n = 4). The median time treated on ICU was significantly longer in the intervention group, compared to the control group (intervention: 14 days (13–17) vs. control: 6 days (1–13), $p = 0.028$). As one patient was not admitted due to COVID-19, we also performed a corrected comparison, which included only patients who were admitted to ICU due to COVID-19 (Table 5)).

Table 5. Odds ratio of COVID-19-related complications.

COVID-19-Related Complications	Odds Ratio	95% Confidence Interval Lower Level	95% Confidence Interval Upper Level	p-Value
ICU Admission	1.371	0.232	4.319	1.00
ICU Admission COVID-19-related [a]	1.371	0.286	6.576	0.693
Intubation	4.343	0.463	40.749	0.199
Other complications	1.542	0.243	9.776	0.646
Death	0.487	0.042	5.601	0.564

[a] One patient was excluded from the corrected analysis, as the reason for ICU admission was not COVID-19-related.

In total, five patients were intubated, without a difference between the groups (intervention: 10.3% (n = 4) vs. control: 2.6% (n = 1); OR 4.343, 95% CI: 0.463–40.749). Overall, three participants died during this trial (intervention: 2.6% (n = 1) vs. control: 5.1% (n = 2), $p = 1.00$; OR 0.487, 95 CI: 0.042–5.601). Other COVID-19-related complications were reported in 5 out of 78 patients (intervention: 7.7% (n = 3) vs. control: 5.1% (n = 2), $p = 1.00$; OR 1.542, 95 CI: 0.243–9.776) (Table 4).

At randomization, the two groups did not differ in the mean serum 25(OH)D_3 level (intervention: 31.46 nmol/L (SD $\pm$ 10.95) (range: 11–44) vs. control: 28.54 nmol/L (SD $\pm$ 10.13) (range: 9–45); $p = 0.232$). At discharge, the intervention group had higher mean serum 25(OH)D_3 D level compared to the control group (intervention: 50.95 nmol/L (SD $\pm$ 16.98) (range: 20–83) vs. control: 26.35 nmol/L (SD $\pm$ 8.88) (range: 9–45); $p < 0.001$). The mean increase in serum 25(OH)D_3 was higher in the intervention group, whereas the mean 25(OH)D_3 level in the control group decreased (intervention: +22.81 nmol/L (SD $\pm$ 15.23) (range: −5–48) vs. control: −2.73 nmol/L (SD $\pm$ 10.23) (range: −33–16); $p < 0.001$). (Table 6, Figure 3).

Table 6. Change of serum 25(OH)D from randomization to discharge.

Serum 25(OH)D Level in nmol/L	N Available (Intervention/Control) [a]	Intervention	Control	p-Value
25(OH)D level at randomization	37/39	31.46 ± 10.948 [11–44]	28.54 ± 10.13 [9–45]	0.232
25(OH)D level at discharge	21/26	50.95 ± 16.98 [2–83]	23.62 ± 8.88 [2–61]	<0.001 ***
Change 25(OH)D level	21/26	26.35 ± 8.88 [9–45]	−2.73 ± 10.23 [−33–16]	<0.001 ***
Patients with 25(OH)D level >50 nmol at discharge	21/26	23.1 (9)	2.6 (1)	0.003 ‡ **

Continuous variables: mean ± standard deviation (SD) [range]. Categorical variables: percentages of patients (absolute number of patients per group = n). [a] number of patients with available information. ‡ Fisher's exact test when expected frequencies <5 instead of Chi^2 test for categorical variables. Level of significance: ** < 0.01 *** < 0.001.

At randomization, there was a significant difference in the mean SF-12 mental score between the groups (intervention: 41.759 $\pm$ 15.510 vs. control: 51.656 $\pm$ 8.044, $p = 0.018$). No significant difference in the mean SF-12 physical score was observable at randomization (intervention: 31.39 $\pm$ 10.61 vs. control: 27.85 $\pm$ 11.75, $p = 0.352$). In the first follow-up call 28 days after inclusion, there was no more significant difference in the mean

SF-12 mental score between the groups observed (intervention: 51.09 ± 13.57 vs. control: 51.86 ± 8.82, p = 0.775). The mean SF-12 physical score did not differ significantly at the first follow-up (intervention: 38.73 ± 9.14 vs. control: 43.41 ± 10.35, p = 0.136). In the second follow-up call 90 days after inclusion, again, no significant difference in the mean SF-12 mental score (intervention: 47.58 ± 11.79 vs. control: 52.81 ± 9.49, p = 0.136) or the mean SF-12 physical score (intervention: 45.07 ± 11.72 vs. control: 50.81 ± 5.70, p = 0.136) was observed (Supplementary Table S5, Supplementary Figure S1). The linear mixed effect model showed a significant time-by-treatment interaction for SF-12 physical evaluation (p = 0.023), indicating that the intervention group had a less pronounced increase of the SF-12 physical scores across time (−4.56 per measured time point). On the SF-12 mental score evaluation, no significant time-by-treatment interaction was observed (p = 0.272, 2.75 per measured time point) (Supplementary Table S6).

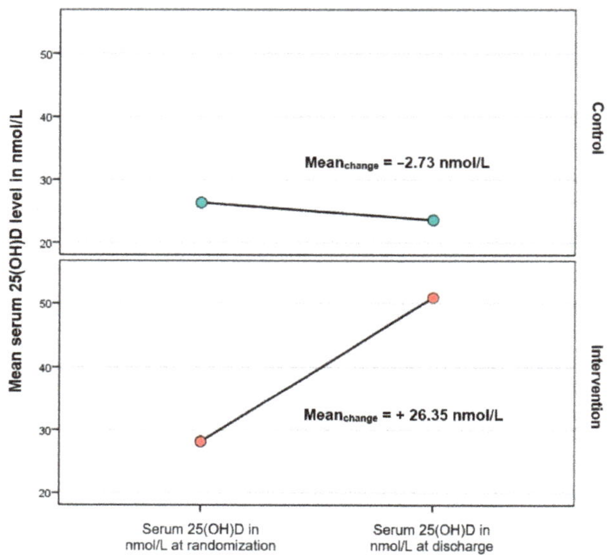

Figure 3. Mean change in serum 25(OH)D from time of randomization.

On day five after randomization, it was observed that significantly more patients reported headache within the intervention group compared to the control group (intervention: 56.3% vs. control: 15%, p = 0.014) (Supplementary Table S7). There was no other significant difference in self-reported symptoms observable at day 5 after randomization. The groups did not differ in terms of vital signs (Supplementary Table S8). Ten days after randomization, there were no more significant differences in terms of self-reported symptoms observed between the groups. In terms of vital signs, the intervention group had a lower systolic blood pressure (p = 0.029), but no difference was observed in other vital signs (Supplementary Table S9, Supplementary Table S10).

Non-significantly fewer adverse events were reported in the intervention group compared to the control group (intervention: 10.3% vs. control: 25.6%, p = 0.138). For serious adverse events, the intervention group had non-significantly more SAEs reported compared to the control group (intervention: 10.3% vs. control: 5.1%, p = 0.675) (Supplementary Table S11). We did not observe a significantly higher mean calcium level in the intervention group compared to the control group (intervention: 2.30 mmol/L (SD ± 0.99) vs. control: 2.22 mmol/L (SD ± 0.063); p= 0.101).

We observed a significant higher mean PTH level in the control group compared to the intervention group (intervention: 2.98 pmol/L (SD ± 1.30) vs. control: 1.60 pmol/L

(SD ± 2.16); $p = 0.007$) and a significant lower mean level of phosphate in the control group compared to the intervention group (intervention: 1.01 mmol/lL(SD ± 0.289) vs. control: 0.86 mmol/L (SD ± 2.66); $p = 0.003$) (Supplementary Table S12).

4. Discussion

We could not observe a difference in the primary endpoint length of hospital stay. The comparison of the length of stay variables between the two groups did not reveal a significant difference that would indicate a beneficial effect of a single high dose vitamin D_3 in addition to standard treatment. The findings of our study are in line with previous results [17,18,22–24,35]. The median length of stay in our population was similar to the findings presented by Murai et al. (2021) [22].

In the post-hoc subgroup analysis, which only included patients with a 25(OH)D_3 deficiency (<25 nmol/L), we observed a non-significant shorter length of stay for the intervention group. There was no difference in the median time from symptom onset to discharge, but in the other length of stay variables ("hospital admission to discharge" and "randomization to discharge"). This trend was already observed by other studies, therefore it might be worth having a closer look at patients with severe 25(OH)D_3 deficiency in bigger studies, as our sub-group was underpowered to detect a clinically significant difference [18,36].

In terms of secondary outcomes, we could not find significant differences in the need of ICU treatment, intubation, death, or other COVID-19 related complications. Those findings indicate that high dose vitamin D_3 had no impact on the course of the disease regarding severity. We observed a significantly longer mean duration in the ICU within the intervention group. In the overall population, there were fewer admissions to the ICU compared to other studies, which was possibly influenced by the fact that our population had fewer co-morbidities and therefore the risk of ICU admission was smaller [22,37]. With our study, we could not support previous findings that high dose vitamin D_3 supplementation reduces ICU admissions [37]. The fact that more patients in the intervention group were intubated compared to the control was not significant and we therefore assume that individual predisposing factors or pre-existing conditions have had a greater influence than the intervention with high doses of vitamin D_3 in patients with 25(OH)D_3 deficiency [5]. This might also contribute to the significantly longer length of ICU treatment in the intervention group. Our study could not confirm the results that fewer patients needed ICU admission or less mechanical ventilation when treated with high dose vitamin D_3 in addition to the best available other therapies [37–39]. However, the number of patients with COVID-19-related complications was very small and therefore the results need to be interpreted with the utmost caution.

We observed a significant increase in the mean serum 25(OH)D_3 level in the intervention group compared to the control group, which resulted in a significant higher mean level at discharge in the intervention group and significantly more patients had serum 25(OH)D_3 levels ≥50 nmol/L in the intervention group. At the same time, there was no clinically relevant increase in calcium or phosphate in the intervention group, which is in line with previous results [24,40–42]. This indicates that a single bolus of high dose vitamin D_3 in addition to daily low dose maintenance therapy is effective to elevate serum 25(OH)D_3 levels quickly. This finding is in line with previous observations that investigated the effect of a single high dose vitamin D_3 with or without daily maintenance dose in various conditions [29,43–45]. Another factor that might have influenced the outcome of our study is the individual response to vitamin D_3 supplementation, as only 23.1% of the participants in the intervention group reached sufficient 25(OH)D_3 serum levels [46].

At time of inclusion, we observed significantly lower SF-12 mental scores in the intervention group than in the control group. This might have led to a bias in the primary outcome variable, since higher hrQoL scores are associated with a reduction of LOHS [47]. The difference between the groups in the SF-12 mental sub-score was not observable anymore 28 days and 90 days after randomization. The increasing number of missing data over time might mediate this observation. Another aspect that should be considered in the discussion of the observation is the range of the SF-12 mental score in our population. In the intervention group, both the range and the variance were larger than in the control group. However, the sample with a completed hrQoL from baseline to day 90 was rather small and therefore possible effects of the two groups did not differ significantly in terms of self-reported symptoms or recorded vital signs. At day five after randomization, we observed that a significantly higher number of patients in the intervention group reported headache compared to the control group, but no differences in other symptoms or vital signs. Ten days after randomization, no more differences were observed in self-reported symptoms. When looking at the vital signs, we observed a significantly lower systolic blood pressure in the intervention group compared to control group, but still the values were in the normal range. Orthostatic hypotension can be caused by longer bedrest or infections [48,49]. Further, only 10 and 11 patients respectively were still hospitalized at day 10 after randomization and therefore the results must be interpreted with caution. Our study could not confirm the findings which reported faster symptom recovery [17,18].

Our results did not show significant differences in the occurrence of adverse and serious adverse events between the intervention group and the control group. This indicates that the intervention with high dose vitamin D_3 is safe and not associated with more adverse or serious adverse events.

Our study population showed similar characteristics at the time of randomization such as age, median time from symptom onset to hospitalization, laboratory parameters, or 25(OH)D_3 level compared to other trials, except the sex distribution, as our study had fewer female participants [17,18,21–24,35,40,41,50–53]. In the early stage of the pandemic in Switzerland, it was observed that the majority of hospitalized patients were male (female: 37% vs. male: 63%), which might explain the lower number of females in our study [54]. In terms of hematological and blood chemistry parameters, we observed several values that were above or beyond the normal value range, which can be explained by viral infections, including SARS-CoV-2, medication, systemic disorders, or malnutrition [55–62]. The intervention group showed significantly higher baseline CRP values than the control group. This difference might imply that patients in the intervention group were suffering from a more severe disease compared to the control group, as a higher CRP could be an early sign for more severe COVID-19 [63,64]. However, since there were no significant differences in baseline respiratory status suggesting a more severe disease in the intervention group (respiratory rate, SpO2, dyspnea, O2-supplementation), the authors cannot conclude if this imbalance had an effect on the overall results of this study.

Our population had a lower frequency of comorbidities reported compared to other studies, namely arterial hypertension and other CVDs [22,24,35,53,65,66], which might influenced the overall outcome.

The time between symptom onset and randomization was rather long with more than 7 days in both groups, which might have had a influence on the primary outcome, as already reported by Murai and colleagues [22].

Strengths and Limitations

Among the strengths of our study are the study design as a randomized, controlled double blind multicenter trial, that the groups did not differ significantly across demographic and clinical variables at time of randomization, and the fact that all patients completed the trial. Limitations for this trial were the small sample size of 80 patients included and 78 analyzed, and missing values in some laboratory parameters at randomization and discharge and the health-related quality of life in the follow-up period. Further, we did not assess COVID-19 severity with a validated scale. In addition, we did not include outpatients in this trial and the period between symptom onset and randomization was rather long.

5. Conclusions

In this trial, we could not confirm previous findings that a bolus of high dose vitamin D_3 had a beneficial effect on the investigated outcomes patients with COVID-19 and 25(OH)D_3 deficiency, except for the elevation of the serum 25(OH)D_3 level. In the subgroup analysis, we observed a non-significant difference in median LOS, which might indicate that patients with severe 25(OH)D_3 deficiency would benefit more from the high dose intervention compared to the overall 25(OH)D_3 deficient population in our study [36,67,68]. Whether this finding was due to the beneficial effect of vitamin D_3 or due to chance remains unclear, since the subgroup analysis was underpowered, and the assumption needs further investigation for verification. In further studies focusing on patients with severe 25(OH)D_3 deficiency, non-hospitalized patients or patients in nursing homes should be included and the outcome of fewer hospital admissions needed should be added. We could show that a single high dose vitamin D_3 in addition to standard vitamin D_3 is not associated with adverse or serious adverse events, and therefore a safe and effective way to elevate the serum 25(OH)D_3 level quickly, whereas daily small dosages of vitamin D_3 supplementation did not elevate the serum 25(OH)D_3 level sufficiently.

6. Participant Privacy and Confidentiality

All individual medical information that was obtained during this trial is considered confidential and disclosure to third parties is prohibited. Only ethics committees or regulatory authorities are allowed to have access to source data within the framework of an audit of the study. The confidentiality is further ensured by utilizing subject identification numbers for data record in the database.

Supplementary Materials: The following supporting information can be downloaded at: https://www.mdpi.com/article/10.3390/biomedicines11051277/s1, Table S1: Visit procedure and collected information for each patient at different time points during the study. Table S2: Co-Morbidities at time of randomization. Table S3: Concomitant medication during hospitalization. Table S4. Robust linear regression results presenting the effect of group. Intervention vs. Control on time to discharge (from randomization) (univariable and adjusted for COVID-19 risk factors, prognostic imbalances, and center effects. Table S5: Mean SF-12 mental and SF-12 physical scores at randomization, 28 days and 90 days after randomization. Table S6: Mixed-effects linear regression results on SF-12 mental and physical outcome with a time and group interaction term. Table S7: Frequency of self- reported symptoms at day 5 after randomization. Table S8: Mean vital signs at day 5 after randomization. Table S9: Frequency of self- reported symptoms at day 10 after randomization. Table S10: Mean vital signs at day 10 after randomization. Table S11: Frequency of adverse and serious adverse events occurred during the trial per study group. Table S12: Levels of calcium, PTH and phosphorus at time of discharge. Figure S1: Group differences in changes in SF-12 mental (top) and SF-12 physical (bottom) across time (randomization = 0m, 28 days = 1 m, 3 months = 3 m). The mean and 95% confidence interval for the observed mean values are displayed.

Author Contributions: Conceptualization, J.D.L., M.B. (Maria Boesing), G.L.-C., M.B. (Michael Braendle), P.H., J.M., A.B.L.-T., P.S. and F.J.; methodology, M.B. (Maria Boesing) and F.J.; formal analysis, S.G. and F.J.; investigation, M.B. (Maria Boesing), M.B. (Michael Braendle), N.B., V.G. and A.R.; resources, J.D.L.; data curation, K.A., V.G. and A.R.; writing—original draft preparation, F.J.; writing—review and editing, M.B. (Maria Boesing), G.L.-C., K.A., N.B., S.G., V.G., A.B.L.-T., P.H., J.M., A.R., P.S., M.B. (Michael Braendle) and J.D.L.; visualization, S.G. and F.J.; supervision, J.D.L.; project administration, F.J. and K.A.; funding acquisition, F.J. and J.D.L. All authors have read and agreed to the published version of the manuscript.

Funding: This research was funded by SWICA AG Switzerland, Spendensitiftung Bank Vontobel, Switzerland, SWF Stiftung für wissenschaftliche Forschung, Switzerland, and Gebro Pharma AG, Switzerland.

Institutional Review Board Statement: The study was approved by the local ethical committees of northwestern and eastern Switzerland on 23 September 2020 and 19 October 2020 (Project ID: 2020-01401) and Swissmedic on the 19 November 2020 (Project ID: 2020DR33170). The study was registered on an international and a national clinical trials registry platform with the registration number NCT04525820.

Informed Consent Statement: Written informed consent was obtained from all subjects involved in the study.

Data Availability Statement: The data presented in this study are available on request from the corresponding author. The data are not publicly available due to privacy policy and ethical statement.

Conflicts of Interest: The authors declare no conflict of interest, and the funders had no role in the design of the study; in the collection, analyses, or interpretation of data; in the writing of the manuscript; or in the decision to publish the results.

References

1. Tazerji, S.S.; Shahabinejad, F.; Tokasi, M.; Rad, M.A.; Khan, M.S.; Safdar, M.; Filipiak, K.J.; Szarpak, L.; Dzieciatkowski, T.; Jurgie J.; et al. Global data analysis and risk factors associated with morbidity and mortality of COVID-19. *Gene Rep.* **2022**, *26*, 10150 [CrossRef] [PubMed]
2. Xiang, L.; Ma, S.; Yu, L.; Wang, W.; Yin, Z. Modeling the Global Dynamic Contagion of COVID-19. *Front. Public Health* **2022** *9*, 809987. [CrossRef]
3. Zhou, H.; Yang, J.; Zhou, C.; Chen, B.; Fang, H.; Chen, S.; Zhang, X.; Wang, L.; Zhang, L. A Review of SARS-CoV2: Compare With SARS-CoV and MERS-CoV. *Front. Med.* **2021**, *8*, 628370. [CrossRef] [PubMed]
4. Cava, E.; Neri, B.; Carbonelli, M.G.; Riso, S.; Carbone, S. Obesity pandemic during COVID-19 outbreak: Narrative review an future considerations. *Clin. Nutr. Edinb. Scotl.* **2021**, *40*, 1637–1643. [CrossRef] [PubMed]
5. Gao, Y.; Ding, M.; Dong, X.; Zhang, J.; Kursat Azkur, A.; Azkur, D.; Gan, H.; Sun, Y.; Fu, W.; Li, W.; et al. Risk factors for sever and critically ill COVID-19 patients: A review. *Allergy* **2021**, *76*, 428–455. [CrossRef]
6. Wilk-Sledziewska, K.; Sielatycki, P.J.; Uscinska, N.; Bujno, E.; Rosolowski, M.; Kakareko, K.; Sledziewski, R.; Rydzewsk Rosolowska, A.; Hryszko, T.; Zbroch, E. The Impact of Cardiovascular Risk Factors on the Course of COVID-19. *J. Clin. Me* **2022**, *11*, 2250. [CrossRef]
7. Parohan, M.; Yaghoubi, S.; Seraji, A.; Javanbakht, M.H.; Sarraf, P.; Djalali, M. Risk factors for mortality in patients wit Coronavirus disease 2019 (COVID-19) infection: A systematic review and meta-analysis of observational studies. *Aging Mal* **2020**, *23*, 1416–1424. [CrossRef]
8. Dessie, Z.G.; Zewotir, T. Mortality-related risk factors of COVID-19: A systematic review and meta-analysis of 42 studies an 423,117 patients. *BMC Infect. Dis.* **2021**, *21*, 855. [CrossRef]
9. Jordan, T.; Siuka, D.; Rotovnik, N.K.; Pfeifer, M. COVID-19 and Vitamin D- a Systematic Review. *Zdr. Varst.* **2022**, *61*, 124–13 [CrossRef] [PubMed]
10. Charoenngam, N.; Holick, M.F. Immunologic Effects of Vitamin D on Human Health and Disease. *Nutrients* **2020**, *12*, 209 [CrossRef]
11. Cannell, J.J.; Vieth, R.; Umhau, J.C.; Holick, M.F.; Grant, W.B.; Madronich, S.; Garland, C.F.; Giovannucci, E. Epidemic influenz and vitamin D. *Epidemiol. Infect.* **2006**, *134*, 1129–1140. [CrossRef]
12. Jolliffe, D.A.; Griffiths, C.J.; Martineau, A.R. Vitamin D in the prevention of acute respiratory infection: Systematic review o clinical studies. *J. Steroid Biochem. Mol. Biol.* **2013**, *136*, 321–329. [CrossRef]
13. Hansdottir, S.; Monick, M.M.; Hinde, S.L.; Lovan, N.; Look, D.C.; Hunninghake, G.W. Respiratory epithelial cells convert inactiv vitamin D to its active form: Potential effects on host defense. *J. Immunol.* **2008**, *181*, 7090–7099. [CrossRef]
14. Olliver, M.; Spelmink, L.; Hiew, J.; Meyer-Hoffert, U.; Henriques-Normark, B.; Bergman, P. Immunomodulatory Effects o Vitamin D on Innate and Adaptive Immune Responses to Streptococcus pneumoniae. *J. Infect. Dis.* **2013**, *208*, 1474–148 [CrossRef] [PubMed]
15. Greiller, C.L.; Martineau, A.R. Modulation of the immune response to respiratory viruses by vitamin D. *Nutrients* **2015**, 4240–4270. [CrossRef]
16. Park, S.; Lee, M.G.; Hong, S.-B.; Lim, C.-M.; Koh, Y.; Huh, J.W. Effect of vitamin D deficiency in Korean patients with acut respiratory distress syndrome. *Korean J. Intern. Med.* **2018**, *33*, 1129–1136. [CrossRef]
17. Sabico, S.; Enani, M.A.; Sheshah, E.; Aljohani, N.J.; Aldisi, D.A.; Alotaibi, N.H.; Alshingetti, N.; Alomar, S.Y.; Alnaami, A.M Amer, O.E.; et al. Effects of a 2-Week 5000 IU versus 1000 IU Vitamin D3 Supplementation on Recovery of Symptoms in Patien with Mild to Moderate COVID-19: A Randomized Clinical Trial. *Nutrients* **2021**, *13*, 2170. [CrossRef]
18. Maghbooli, Z.; Sahraian, M.A.; Jamalimoghadamsiahkali, S.; Asadi, A.; Zarei, A.; Zendehdel, A.; Varzandi, T.; Mohammadnab S.; Alijani, N.; Karimi, M.; et al. Treatment with 25-Hydroxyvitamin D(3) (Calcifediol) Is Associated With a Reduction in th Blood Neutrophil-to-Lymphocyte Ratio Marker of Disease Severity in Hospitalized Patients With COVID-19: A Pilot Multicente Randomized, Placebo-Controlled, Double-Blinded Clinical Trial. *Endocr. Pract. Off. J. Am. Coll. Endocrinol. Am. Assoc. Cli Endocrinol.* **2021**, *27*, 1242–1251. [CrossRef]

19. Karonova, T.L.; Andreeva, A.T.; Golovatuk, K.A.; Bykova, E.S.; Simanenkova, A.V.; Vashukova, M.A.; Grant, W.B.; Shlyakhto, E.V. Low 25(OH)D Level Is Associated with Severe Course and Poor Prognosis in COVID-19. *Nutrients* **2021**, *13*, 3021. [CrossRef] [PubMed]
20. Rastogi, A.; Bhansali, A.; Khare, N.; Suri, V.; Yaddanapudi, N.; Sachdeva, N.; Puri, G.D.; Malhotra, P. Short term, high-dose vitamin D supplementation for COVID-19 disease: A randomised, placebo-controlled, study (SHADE study). *Postgrad. Med. J.* **2022**, *98*, 87. [CrossRef]
21. Sulli, A.; Gotelli, E.; Casabella, A.; Paolino, S.; Pizzorni, C.; Alessandri, E.; Grosso, M.; Ferone, D.; Smith, V.; Cutolo, M. Vitamin D and Lung Outcomes in Elderly COVID-19 Patients. *Nutrients* **2021**, *13*, 717. [CrossRef]
22. Murai, I.H.; Fernandes, A.L.; Sales, L.P.; Pinto, A.J.; Goessler, K.F.; Duran, C.S.C.; Silva, C.B.R.; Franco, A.S.; Macedo, M.B.; Dalmolin, H.H.H.; et al. Effect of a Single High Dose of Vitamin D3 on Hospital Length of Stay in Patients With Moderate to Severe COVID-19: A Randomized Clinical Trial. *JAMA* **2021**, *325*, 1053–1060. [CrossRef] [PubMed]
23. Elamir, Y.M.; Amir, H.; Lim, S.; Rana, Y.P.; Lopez, C.G.; Feliciano, N.V.; Omar, A.; Grist, W.P.; Via, M.A. A randomized pilot study using calcitriol in hospitalized COVID-19 patients. *Bone* **2022**, *154*, 116175. [CrossRef]
24. Cannata-Andía, J.B.; Díaz-Sottolano, A.; Fernández, P.; Palomo-Antequera, C.; Herrero-Puente, P.; Mouzo, R.; Carrillo-López, N.; Panizo, S.; Ibañez, G.H.; Cusumano, C.A.; et al. A single-oral bolus of 100,000 IU of cholecalciferol at hospital admission did not improve outcomes in the COVID-19 disease: The COVID-VIT-D-a randomised multicentre international clinical trial. *BMC Med.* **2022**, *20*, 83. [CrossRef] [PubMed]
25. Gao, Z.; Xie, J.; Li, C.; Liu, L.; Yang, Y. High Dose Vitamin D3 Supplementation Is Not Associated With Lower Mortality in Critically Ill Patients: A Meta-Analysis of Randomized Control Trials. *Front. Nutr.* **2022**, *9*, 762316. [CrossRef]
26. Malihi, Z.; Lawes, C.M.M.; Wu, Z.; Huang, Y.; Waayer, D.; Toop, L.; Khaw, K.-T.; Camargo, C.A., Jr.; Scragg, R. Monthly high-dose vitamin D supplementation does not increase kidney stone risk or serum calcium: Results from a randomized controlled trial. *Am. J. Clin. Nutr.* **2019**, *109*, 1578. [CrossRef]
27. Amrein, K.; Sourij, H.; Wagner, G.; Holl, A.; Pieber, T.R.; Smolle, K.H.; Stojakovic, T.; Schnedl, C.; Dobnig, H. Short-term effects of high-dose oral vitamin D3 in critically ill vitamin D deficient patients: A randomized, double-blind, placebo-controlled pilot study. *Crit. Care* **2011**, *15*, R104. [CrossRef] [PubMed]
28. Apaydin, M.; Can, A.G.; Kizilgul, M.; Beysel, S.; Kan, S.; Caliskan, M.; Demirci, T.; Ozcelik, O.; Ozbek, M.; Cakal, E. The effects of single high-dose or daily low-dosage oral colecalciferol treatment on vitamin D levels and muscle strength in postmenopausal women. *BMC Endocr. Disord.* **2018**, *18*, 48. [CrossRef] [PubMed]
29. Ganmaa, D.; Munkhzul, B.; Fawzi, W.; Spiegelman, D.; Willett, W.C.; Bayasgalan, P.; Baasansuren, E.; Buyankhishig, B.; Oyun-Erdene, S.; Jolliffe, D.A.; et al. High-Dose Vitamin D(3) during Tuberculosis Treatment in Mongolia. A Randomized Controlled Trial. *Am. J. Respir. Crit. Care Med.* **2017**, *196*, 628–637. [CrossRef]
30. Ataide, F.L.; Carvalho Bastos, L.M.; Vicente Matias, M.F.; Skare, T.L.; Freire de Carvalho, J. Safety and effectiveness of vitamin D mega-dose: A systematic review. *Clin. Nutr. ESPEN* **2021**, *46*, 115–120. [CrossRef]
31. Ramasamy, I. Vitamin D Metabolism and Guidelines for Vitamin D Supplementation. *Clin. Biochem. Rev.* **2020**, *41*, 103–126. [CrossRef]
32. ICHH Guideline. International Conference on Harmonisation of Technical Requirements for Pharmaceuticals for Human Use (ICH) Integrated addendum to ICH E6 (R1): Guideline for good clinical practice E6 (R2). *Curr. Step* **2015**, *2*, 1–60.
33. Guan, W.-J.; Ni, Z.-Y.; Hu, Y.; Liang, W.-H.; Ou, C.-Q.; He, J.-X.; Liu, L.; Shan, H.; Lei, C.-L.; Hui, D.S.C.; et al. Clinical Characteristics of Coronavirus Disease 2019 in China. *N. Engl. J. Med.* **2020**, *382*, 1708–1720. [CrossRef]
34. Jaun, F.; Boesing, M.; Lüthi-Corridori, G.; Abig, K.; Makhdoomi, A.; Bloch, N.; Lins, C.; Raess, A.; Grillmayr, V.; Haas, P.; et al. High-dose vitamin D substitution in patients with COVID-19: Study protocol for a randomized, double-blind, placebo-controlled, multi-center study—VitCov Trial. *Trials* **2022**, *23*, 114. [CrossRef]
35. Mariani, J.; Antonietti, L.; Tajer, C.; Ferder, L.; Inserra, F.; Sanchez Cunto, M.; Brosio, D.; Ross, F.; Zylberman, M.; López, D.E.; et al. High-dose vitamin D versus placebo to prevent complications in COVID-19 patients: Multicentre randomized controlled clinical trial. *PLoS ONE* **2022**, *17*, e0267918. [CrossRef] [PubMed]
36. De Niet, S.; Trémège, M.; Coffiner, M.; Rousseau, A.-F.; Calmes, D.; Frix, A.-N.; Gester, F.; Delvaux, M.; Dive, A.-F.; Guglielmi, E.; et al. Positive Effects of Vitamin D Supplementation in Patients Hospitalized for COVID-19: A Randomized, Double-Blind, Placebo-Controlled Trial. *Nutrients* **2022**, *14*, 3048. [CrossRef]
37. Entrenas Castillo, M.; Entrenas Costa, L.M.; Vaquero Barrios, J.M.; Alcalá Díaz, J.F.; López Miranda, J.; Bouillon, R.; Quesada Gomez, J.M. Effect of calcifediol treatment and best available therapy versus best available therapy on intensive care unit admission and mortality among patients hospitalized for COVID-19: A pilot randomized clinical study. *J. Steroid Biochem. Mol. Biol.* **2020**, *203*, 105751. [CrossRef] [PubMed]
38. Annweiler, G.; Corvaisier, M.; Gautier, J.; Dubée, V.; Legrand, E.; Sacco, G.; Annweiler, C. Vitamin D Supplementation Associated to Better Survival in Hospitalized Frail Elderly COVID-19 Patients: The GERIA-COVID Quasi-Experimental Study. *Nutrients* **2020**, *12*, 3377. [CrossRef] [PubMed]
39. Asghar, M.S.; Yasmin, F.; Dapke, K.; Shah, S.M.I.; Zafar, M.D.B.; Khan, A.A.; Mohiuddin, O.; Surani, S. Evaluation of Vitamin-D Status and Its Association with Clinical Outcomes Among COVID-19 Patients in Pakistan. *Am. J. Trop. Med. Hyg.* **2022**, *106*, 150–155. [CrossRef] [PubMed]

40. Karonova, T.L.; Golovatyuk, K.A.; Kudryavtsev, I.V.; Chernikova, A.T.; Mikhaylova, A.A.; Aquino, A.D.; Lagutina, D.I.; Zaikova, E.K.; Kalinina, O.V.; Golovkin, A.S.; et al. Effect of Cholecalciferol Supplementation on the Clinical Features and Inflammatory Markers in Hospitalized COVID-19 Patients: A Randomized, Open-Label, Single-Center Study. *Nutrients* **2022**, *14*, 2602. [CrossRef]
41. Torres, M.; Casado, G.; Vigón, L.; Rodríguez-Mora, S.; Mateos, E.; Ramos-Martín, F.; López-Wolf, D.; Sanz-Moreno, J.; Ryan-Murua, P.; Taboada-Martínez, M.L.; et al. Changes in the immune response against SARS-CoV-2 in individuals with severe COVID-19 treated with high dose of vitamin D. *Biomed. Pharmacother.* **2022**, *150*, 112965. [CrossRef]
42. Cervero, M.; López-Wolf, D.; Casado, G.; Novella-Mena, M.; Ryan-Murua, P.; Taboada-Martínez, M.L.; Rodríguez-Mora, S.; Vigón, L.; Coiras, M.; Torres, M. Beneficial Effect of Short-Term Supplementation of High Dose of Vitamin D3 in Hospitalized Patients With COVID-19: A Multicenter, Single-Blinded, Prospective Randomized Pilot Clinical Trial. *Front. Pharmacol.* **2022**, *13*, 863587. [CrossRef]
43. Alsufiani, H.M.; AlGhamdi, S.A.; AlShaibi, H.F.; Khoja, S.O.; Saif, S.F.; Carlberg, C. A Single Vitamin D3 Bolus Supplementation Improves Vitamin D Status and Reduces Proinflammatory Cytokines in Healthy Females. *Nutrients* **2022**, *14*, 3963. [CrossRef]
44. Martineau, A.R.; MacLaughlin, B.D.; Hooper, R.L.; Barnes, N.C.; Jolliffe, D.A.; Greiller, C.L.; Kilpin, K.; McLaughlin, D.; Fletcher, G.; Mein, C.A.; et al. Double-blind randomised placebo-controlled trial of bolus-dose vitamin D_3 supplementation in adults with asthma (ViDiAs). *Thorax* **2015**, *70*, 451. [CrossRef]
45. Hiremath, G.; Cettomai, D.; Baynes, M.; Ratchford, J.; Newsome, S.; Harrison, D.; Kerr, D.; Greenberg, B.; Calabresi, P. Vitamin D status and effect of low-dose cholecalciferol and high-dose ergocalciferol supplementation in multiple sclerosis. *Mult. Scler. J.* **2009**, *15*, 735–740. [CrossRef]
46. Carlberg, C.; Haq, A. The concept of the personal vitamin D response index. *Vitam. Defic. Hum. Health* **2018**, *175*, 12–17. [CrossRef] [PubMed]
47. Lin, C.-F.; Huang, Y.-H.; Ju, L.-Y.; Weng, S.-C.; Lee, Y.-S.; Chou, Y.-Y.; Lin, C.-S.; Lin, S.-Y. Health-Related Quality of Life Measured by EQ-5D in Relation to Hospital Stay and Readmission in Elderly Patients Hospitalized for Acute Illness. *Int. J. Environ. Res. Public. Health* **2020**, *17*, 5333. [CrossRef] [PubMed]
48. Noh, S.-M.; Kang, H.G.; Kim, B.J. Syncope after Influenza Virus Infection. *J. Korean Med. Sci.* **2020**, *35*, 1145989. [CrossRef]
49. Marando, M.; Fusi-Schmidhauser, T.; Tamburello, A.; Grazioli Gauthier, L.; Rigamonti, E.; Argentieri, G.; Puligheddu, C.; Pagnamenta, A.; Valenti, A.; Pons, M.; et al. 1-year radiological, functional and quality-of-life outcomes in patients with SARS-CoV-2 pneumonia—A prospective observational study. *NPJ Prim. Care Respir. Med.* **2022**, *32*, 8. [CrossRef] [PubMed]
50. Annweiler, C.; Beaudenon, M.; Simon, R.; Guenet, M.; Otekpo, M.; Célarier, T.; Gautier, J. GERIA-COVID study group Vitamin D supplementation prior to or during COVID-19 associated with better 3-month survival in geriatric patients: Extension phase of the GERIA-COVID study. *J. Steroid Biochem. Mol. Biol.* **2021**, *213*, 105958. [CrossRef]
51. AlSafar, H.; Grant, W.B.; Hijazi, R.; Uddin, M.; Alkaabi, N.; Tay, G.; Mahboub, B.; Al Anouti, F. COVID-19 Disease Severity and Death in Relation to Vitamin D Status among SARS-CoV-2-Positive UAE Residents. *Nutrients* **2021**, *13*, 1714. [CrossRef] [PubMed]
52. Annweiler, C.; Beaudenon, M.; Gautier, J.; Gonsard, J.; Boucher, S.; Chapelet, G.; Darsonval, A.; Fougère, B.; Guérin, O.; Houvet, M.; et al. High-dose versus standard-dose vitamin D supplementation in older adults with COVID-19 (COVIT-TRIAL): A multicenter, open-label, randomized controlled superiority trial. *PLoS Med.* **2022**, *19*, e1003999. [CrossRef]
53. Saponaro, F.; Franzini, M.; Okoye, C.; Antognoli, R.; Campi, B.; Scalese, M.; Neri, T.; Carrozzi, L.; Monzani, F.; Zucchi, R.; et al. Is There a Crucial Link Between Vitamin D Status and Inflammatory Response in Patients with COVID-19? *Front. Immunol.* **2022**, *12*, 745713. [CrossRef]
54. Gregoriano, C.; Koch, D.; Haubitz, S.; Conen, A.; Fux, C.A.; Mueller, B.; Bernasconi, L.; Hammerer-Lercher, A.; Oberle, M.; Burgermeister, S. Characteristics, predictors and outcomes among 99 patients hospitalised with COVID-19 in a tertiary care centre in Switzerland: An observational analysis. *Swiss Med. Wkly.* **2020**, *150*, w20316. [CrossRef]
55. Tahir, N.; Zahra, F. Neutrophilia. In *StatPearls [Internet]*; StatPearls Publishing: Treasure Island, FL, USA, 2022.
56. Rosales, C. Neutrophil: A Cell with Many Roles in Inflammation or Several Cell Types? *Front. Physiol.* **2018**, *9*, 113. [CrossRef]
57. Mayadas, T.N.; Cullere, X.; Lowell, C.A. The Multifaceted Functions of Neutrophils. *Annu. Rev. Pathol. Mech. Dis.* **2014**, *9*, 181–218. [CrossRef]
58. Brass, D.; Mckay, P.; Scott, F. Investigating an incidental finding of lymphopenia. *BMJ* **2014**, *348*, g1721. [CrossRef]
59. Vijayakumar, S.; Viswanathan, S.; Aghoram, R. Idiopathic CD4 lymphocytopenia: Current insights. *Immunotargets Ther.* **2020**, *9*, 79. [CrossRef]
60. Hall, P.; Cash, J. What is the real function of the liver 'function'tests? *Ulster Med. J.* **2012**, *81*, 30. [PubMed]
61. Luo, M.; Ballester, M.P.; Soffientini, U.; Jalan, R.; Mehta, G. SARS-CoV-2 infection and liver involvement. *Hepatol. Int.* **2022**, *16*, 755–774. [CrossRef] [PubMed]
62. Clark, R.; Waters, B.; Stanfill, A.G. Elevated liver function tests in COVID-19: Causes, clinical evidence, and potential treatments *Nurse Pract.* **2021**, *46*, 21–26. [CrossRef]
63. Ali, N. Elevated level of C-reactive protein may be an early marker to predict risk for severity of COVID-19. *J. Med. Virol.* **2020**, *92*, 2409–2411. [CrossRef]
64. Luan, Y.; Yin, C.; Yao, Y. Update Advances on C-Reactive Protein in COVID-19 and Other Viral Infections. *Front. Immunol.* **2021**, *12*, 720363. [CrossRef] [PubMed]

55. Al-Jarallah, M.; Rajan, R.; Dashti, R.; Al Saber, A.; Pan, J.; Zhanna, K.D.; Abdelnaby, H.; Aboelhassan, W.; Almutairi, F.; Abdullah, M.; et al. In-hospital mortality in SARS-CoV-2 stratified by serum 25-hydroxy-vitamin D levels: A retrospective study. *J. Med. Virol.* **2021**, *93*, 5880–5885. [CrossRef]
56. Davoudi, A.; Najafi, N.; Aarabi, M.; Tayebi, A.; Nikaeen, R.; Izadyar, H.; Salar, Z.; Delavarian, L.; Vaseghi, N.; Daftarian, Z.; et al. Lack of association between vitamin D insufficiency and clinical outcomes of patients with COVID-19 infection. *BMC Infect. Dis.* **2021**, *21*, 450. [CrossRef] [PubMed]
57. da Rocha, A.P.; Atallah, A.N.; Aldrighi, J.M.; Pires, A.L.R.; Dos Santos Puga, M.E.; Pinto, A.C.P.N. Insufficient evidence for vitamin D use in COVID-19: A rapid systematic review. *Int. J. Clin. Pract.* **2021**, *75*, e14649. [CrossRef] [PubMed]
58. Malinverni, S.; Ochogavia, Q.; Lecrenier, S.; Scorpinniti, M.; Preiser, J.-C.; Cotton, F.; Mols, P.; Bartiaux, M. Severe vitamin D deficiency in patients admitted to the emergency department with severe sepsis is associated with an increased 90-day mortality. *Emerg. Med. J.* **2022**, *40*, 36–41. [CrossRef]

biomedicines

MDPI

Systematic Review

Immunomodulatory Role of Vitamin D on Gut Microbiome in Children

Anika Tabassum [1], Adli Ali [2], Farah Dayana Zahedi [3] and Noor Akmal Shareela Ismail [1,*]

1 Department of Biochemistry, Faculty of Medicine, Universiti Kebangsaan Malaysia, Kuala Lumpur 56000, Malaysia; p124173@siswa.ukm.edu.my
2 Department of Pediatric, Faculty of Medicine, Universiti Kebangsaan Malaysia, Kuala Lumpur 56000, Malaysia; adli.ali@ppukm.ukm.edu.my
3 Department of Otorhinolaryngology, Faculty of Medicine, Universiti Kebangsaan Malaysia, Kuala Lumpur 56000, Malaysia; farahdayana@ukm.edu.my
* Correspondence: nasismail@ukm.edu.my; Tel.: +60-391459552

Abstract: Vitamin D plays a role in regulating the immune system and can be linked to the alteration of the gut microbiome, which leads to several immunological diseases. This systematic review aims to explore the relationship between Vitamin D and children's gut microbiome, as well as its impact towards the immune system. We have systematically collated relevant studies from different databases concerning changes in the gut microbiome of children from infants to 18 years old associated with Vitamin D and the immunological pathways. The studies utilized 16S rRNA sequencing analysis of fecal matter with or without Vitamin D supplementation and Vitamin D levels. Ten studies were selected for the review, among which eight studies showed significant alterations in the gut microbiome related to Vitamin D supplementation or Vitamin D levels. The taxa of the phylum Actinobacteria, Bacteroidetes, Firmicutes, and Proteobacteria are the most altered in these studies. The alteration of the taxa alters the Th1 and Th2 pathways and changes the immune response. We will discuss how Vitamin D may contribute to the activation of immune pathways via its effects on intestinal barrier function, microbiome composition, and/or direct effects on immune responses. In conclusion, the studies examined in this review have provided evidence that Vitamin D levels may have an impact on the composition of children's gut microbiomes.

Keywords: Vitamin D; gut microbiome; immune system; taxa; children

1. Introduction

The human gut is a reservoir to a rich and diverse community of microbes [1]. The population of microorganisms inhabiting the gut is known as the gut microbiota, and the genome of those microbes refers to the gut microbiome [2]. The gut microbiota plays a vital part in our well-being and illnesses by metabolizing energy [3], producing short chain fatty acids (SCFA) [4] and vitamins [5] while regulating the immune system [6]. The genomic composition and diversity of the human gut is also responsible for childhood development, which includes the development of the immune system and related diseases [7].

The development of the microflora begins at birth, which is passed down from the mother and continues to develop throughout life but becomes comparatively stable when one is older [8]. The infant microbiome is the most susceptible to change, which makes prenatal and the first few years of microbiome development the most vital. The initial microbiota can regulate the infant epithelial gene expression, which determines the adult microbiome balance at a later age. Thus, reviewing the microbiome at an earlier age could control the immune system, as it is the initial stimulus at infancy. The childhood microbiome has various antigens, such as lipoproteins and polysaccharides, which help establish the immune system. Different compositions of microbes will form the immune system differently and it will also affect the immune system related diseases in that individual in childhood and in later life [9].

Immunological diseases have been associated with dysbiosis in children, in which the microbiome composition has been altered [10,11]. Diseases, such as Inflammatory Bowel Diseases (IBD) [12], allergic diseases [10], pediatric multiple sclerosis [13], Type 1 diabetes (T1D) [11], and asthma [14], are commonly associated with dysbiosis in children. Thus, the development of a healthy pediatric gut microbiome is important to prevent diseases and maintain better health. Many factors, such as children's diets, antibiotic exposure, breast milk, or formula feeding of infants, can shape the gut microbiota composition and their function [15–18]. One of the most discussed factors affecting the gut microbiota has been the role of Vitamin D [19]. In past years, metagenomics has enabled researchers to take a comprehensive look into the human gut microbiome, which allows us to observe the shift in the composition of microbiomes associated with Vitamin D supplementation [20]. These analyses help us to elucidate the composition and diversity of the microbes, including their evenness, richness, and functional capacity. Several studies included high throughput 16S rRNA sequencing of the stool sample to successfully analyze the composition and detailed functional gene expression profile of the gut microbiome [21,22]. This 16S rRNA technique uses 16S rRNA as a marker to sequence genes, which has conserved hypervariable regions that can specifically identify the bacteria [23]. Studying the gut microbiome profile during childhood plays an important role in identifying the abundant bacteria and determining the effect of dysbiosis in children's health and immunological diseases in relation to Vitamin D [24].

Vitamin D levels have been associated with improved health and a reduced incidence of immunological diseases in pregnancy and children [25]. Supplementation in deficient children with Vitamin D reduced diseases like asthma, allergic rhinitis, and IBD [26–28]. Additionally, Vitamin D has been suggested to be a key player in altering the composition of the gut microbiome. The immunological and anti-inflammatory role of Vitamin D could lead to the regulation of microbiomes. Studies have shown that Vitamin D regulates several expressions of immune cells [29], including immunoglobulin E (IgE), T-helper 1 (Th-1), and T-helper 2 (Th-2) pro-inflammtory and anti-inflammatory cytokines [30,31], and further maintains the role of the gastrointestinal barrier, which can alter the microbiome [32,33].

The characterization and function of the children's gut microbiome are varied and must be observed, especially concerning the shift in the microbiota with their Vitamin D levels and supplementation among healthy children. There have been limited studies focusing on children's gut microbiome associated with Vitamin D supplementation. Therefore, to investigate the immunomodulatory role of Vitamin D on the gut microbiome, we reviewed clinical and observational studies. This systematic review specifically focused on the prenatal and children's Vitamin D levels or supplementation impact on the gut microbiome of the pediatric population.

2. Materials and Methods

2.1. Study Protocol

We followed the Preferred Reporting Items for Systematic reviews and Meta-Analyses (PRISMA) guideline for systematic reviews to conduct this review. Database search was conducted to find articles from January 1996 to January 2023 from the following databases: (1) PubMed, (2) Medline, (3) Scopus, and (4) Web of Science. Studies were also obtained by looking through reference lists in Google Scholar. Limits were applied during the search, including full-length text articles in English. The terms used for the search were: "Vitamin D and microbiome", "Vitamin D and microbiome and immune", "Vitamin D and microbiome and infection", "Vitamin D and microbiome and inflammation", "Vitamin D and microbiome and profiling", "Vitamin D and microbiome and transcript", and "Vitamin D and microbiome and pediatric", "Vitamin D and microbiome and children". There was no further retrieval of articles needed. No studies were excluded, and no additional analysis was conducted in order to reduce the risk of biasness. To keep the heterogeneity low, the methodology of the studies was kept constant. The detailed search strategy is provided in Appendix A.

2.2. Eligibility Criteria

The abstracts of each article were reviewed initially to check for eligibility. The studies that were included were clinical trials, randomized controlled clinical trials, and cohort studies. The children included were not limited to healthy children. Studies including children ranging from infants to 18 years of age were selected. The selected studies contained a measure of Vitamin D levels, supplementation of Vitamin D, or prenatal supplementation of Vitamin D. The outcome of the studies was recorded as the measure of variability within a sample (alpha diversity) and between samples in microbial composition (beta diversity), species richness, prevalence, and/or (relative) abundance of bacterial taxa. The studies were excluded if they did not analyze the gut microbiome with 16S rRNA gene sequencing. The studies and reviews that were in any language other than English were also excluded.

3. Results

3.1. Study Selection

Across all databases, 2072 articles were obtained, and 82 duplicates were removed. The titles were then skimmed to select 22 articles from PubMed, Medline, Scopus, and WOS, as well as 35 articles from Google Scholar, according to their relevance. Next, the abstract and full articles were studied to exclude the articles that did not fall into the eligibility criteria. Two reviewers (AT, NASI) independently searched for the articles. Twenty-two full-text articles were screened from different databases, and 35 full-text articles were screened from Google Scholar. We selected 10 studies for this review. A flowchart for article identification, screening, and final selection is given below (Figure 1).

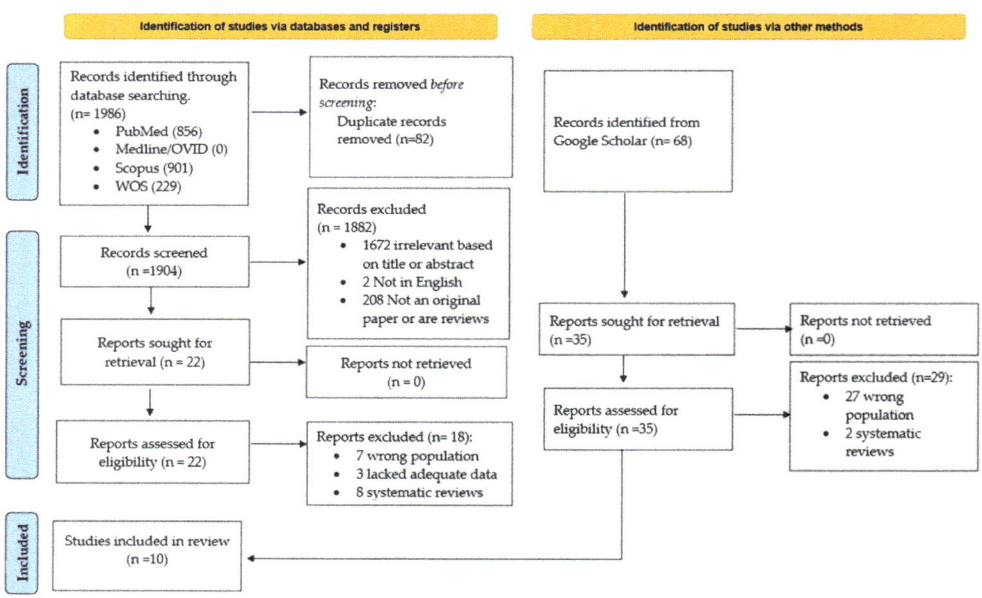

Figure 1. Flowchart to show study selection procedure.

3.2. Study Characteristics

Out of the 10 studies, eight studies included administering Vitamin D supplementation, and two studies included looking into the blood Vitamin D levels. Two studies included administering prenatal Vitamin D supplementation of 400 IU versus 4000 IU [34] and 400 IU versus 2800 IU [35]. One study included administering Vitamin D supplementation of 400 IU or placebo to infants [36]. One study did not include the amount of Vitamin D supplementation to infants [37], while two studies included administering Vitamin D supplementation to both mothers and infants. In one of these two, mothers received a dose of 0 IU versus < 400 IU/day versus ≥ 400 IU/day, while the dosage for infants was not mentioned [38]. In the other, mothers and infants were both were given 400 IU/day. Overall, one study included supplementing high doses of Vitamin D to adolescent girls 50,000 IU/week [39]. One study included providing Vitamin D 400 IU/day or Vitamin D 400 IU/day with Vitamin D 200 or 400 IU/day with another dosage of Vitamin D 400 IU/day or 400 IU/day with 800 IU/day [40]. One out of the 10 studies included Vitamin D levels in 4-to-14-year-old children [41]. One study had maternal and infant vitamin D supplementation of >400 IU/day or <400 IU/day or 400 IU/day [42]. Another study examined the maternal and prenatal Vitamin D levels [43]. The studies are included, and their characteristics are summarized in the table below (Table 1). The primary outcome of the investigations was documented as an abundance of bacterial taxa and alpha or beta diversity of the samples. There were no requirements for any data conversion. In all studies, the gut microbiome measurement was conducted using fecal samples. All of these studies utilized 16S rRNA gene sequencing from fecal samples to reduce the variability in the microbiome measurement.

Table 1. Characteristics of the studies that use Vitamin D supplementation and its effect on the gut microbiome.

Title	Reference	Objective	Type of Study	Sample Size	Age Range	Lab Findings
Factors influencing the infant gut microbiome at age 3–6 months: findings from the ethnically diverse Vitamin D Antenatal Asthma Reduction Trial (VDAART).	[34]	To ascertain the effects of prenatal and early life factors on the gut microbiome in a sizable, ethnically diverse study population of infants who were enrolled in the Vitamin D Antenatal Asthma Reduction Trial, a clinical trial of Vitamin D supplementation in pregnancy to prevent allergies and asthma in offspring.	Cross-sectional (nested within an RCT)	333	3–6 months old	Lachnobacterium, Lachnospiraceae and C. clostridiales coabundance, and a decrease in Lactococcus, were all associated with Vitamin D levels in cord blood immediately following birth (in utero)
Prenatal dietary supplements influence the infant airway microbiota in a randomized factorial clinical trial	[35]	To determine if providing mothers with n-3 LCPUFA and Vitamin D influences the microbiome of the infant and mother and to clarify if the protective impact of these supplements on asthma can be regulated by the microbiota.	Random-ized clinical trials	580	Prenatal–1-year-old	No significant data

Table 1. *Cont.*

Title	Reference	Objective	Type of Study	Sample Size	Age Range	Lab Findings
Metagenomic analysis of the gut microbiome composition associated with Vitamin D supplementation in Taiwanese infants	[36]	To ascertain the impact of Vitamin D supplementation on the development of the infant gut microbiome in Taiwan, as well as to look into the distribution and diversity of the infant gut microbiota under various feeding regimens, both with and without Vitamin D administration.	Clinical trials	62	0–4 months old	Infants that received Vitamin D supplements showed elevated levels of the genus Lactobacillus, Streptococcus, and Bifidobacterium, as well as a greater F/B ratio. Supplementing with Vitamin D decreased the amount of Bacteroides. The majority of the pathways identified were found to be engaged in various types of antibiotic production, which suggests that microbial communities exist in highly competitive environments, according to the metabolic analysis profiles in breastfed infants with Vitamin D deficiency.
Vitamin D Supplementation in Exclusively Breastfed Infants Is Associated with Alterations in the Fecal Microbiome	[37]	In order to ascertain the relationship between maternal and infant characteristics, infant feeding practises, and the gut microbiota profiles in 3- to 9-month-old infants, we separately analysed infants who were exclusively breastfed and those who were not in order to evaluate the impact of different feeding practises features on the microbiotas of the infants in each group.	Cohort study	191	3–9 months old	Infants who were exclusively breastfed and received Vitamin D supplements saw a decline in hemophilus
Influence of Vitamin D on key bacterial taxa in infant microbiota in the KOALA Birth Cohort Study	[38]	To determine if maternal Vitamin D administration, maternal plasma 25-hydroxyvitamin D concentration, or newborn Vitamin D administration directly affects major bacterial taxa within newborns' microbiome at one month old.	Cohort study	913	1 month old	*Bifidobacterium* spp. decreased when maternal Vitamin D supplementation levels increased. C. difficile levels in breastfed infants with Vitamin D supplements were lower as well.

Table 1. *Cont.*

Title	Reference	Objective	Type of Study	Sample Size	Age Range	Lab Findings
The effects of high doses of Vitamin D on the composition of the gut microbiome of adolescent girls	[39]	To assess the impact of a high-dose Vitamin D supplement on the gut microbiome's composition.	Cohort study	50	12–18 years old	The expression folds of Enterococcus (1.05), Bifidobacterium (1.20), and Firmicutes (1.50) all increase with Vitamin D treatment. While Lactobacillus 24% ($p = 0.006$) and Bacteroidetes 72% ($p < 0.0001$) declined.
Effect of Different Doses of Vitamin D on the Intestinal Flora of Babies with Eczema: An Experimental Study	[40]	To determine whether altering the Vitamin D dosage can restore the immune system and intestinal flora of infants with eczema who typically consume 400 IU of Vitamin D per day.		32	1–12 months	With low Vitamin D dosage compared to no supplementation, Prevotella, species Prevotella copri, and family Prevotellaceae decreased in children with eczema. Children with eczema had higher levels of the bacteria Bacteroides vulgatus, Phascolarctobacterium faecium, order Acidaminococcales, family Acidaminococcaceae, genus Phascolarctobacterium, f-Sutterellaceae, and species Parasutterella excrementihominis with low Vitamin D supplementation compared to no supplementation.
Tipping the Balance: Vitamin D Inadequacy in Children Impacts the Major Gut Bacterial Phyla	[41]	To comprehend the incidence, causes of it, and effects of inadequate Vitamin D in Qatari children.	Cohort study	64	4–14 years old	Prevotella 9 and Bacteroidetes increased in Vitamin D-deficient children (50 nmol/L), but Firmicutes, Bacteroides, Alistipes, B/P ratio, and F/B reduced. Prevotella dislens, Prevotella bivia, Bacteroides massillensis, Alistipes marseille, and Alistipes indistinctus were among the species that were more prevalent. Bacteroides and Alistipes, including Alistipes finegoldii, Alistipes sp. AL1, Alistipes sp. N15 MGS-157, Bacteroides caceae, Bacteroides eggerthii, and Bacteroides plebeius, were more prevalent in Vitamin D non-deficient (>50 nmol/L) individuals.

Table 1. *Cont.*

Title	Reference	Objective	Type of Study	Sample Size	Age Range	Lab Findings
Vitamin D supplementa-tion in pregnancy and early infancy in relation to gut microbiota composition and C. difficile colonization: Implications for viral respiratory infections	[42]	In 1157 mother–infant couples participating in the CHILD (Canadian Healthy Infant Longitudinal Development) Cohort Study from 2009 to 2012, the connection between maternal and newborn Vitamin D supplementation, infant gut microbiota makeup, and Clostridioides difficile colonisation was examined.	Cohort study	1157	3–4 months old	Genus Megamonas declines in infants receiving Vitamin D supplements. In breastfed newborns, Bilophila (-) and Lachnospiraceae decrease with maternal Vitamin D administration, but Haemophilus rises. Once breastfeeding status and other variables were taken into account, neither infant nor maternal Vitamin D supplementation was linked to C. difficile colonisation. Infant colonisation with C. difficile was less likely when mothers drank milk that was supplemented with Vitamin D.
Maternal and cord blood Vitamin D level, and the infant gut microbiota in a birth cohort study	[43]	To investigate relationships between newborn gut microbiota and plasma 25[OH]D levels throughout pregnancy or at birth (cord blood).	Cohort study	499	1–6 months old	Acinetobacter and Corynebacterium increased with maternal and cord Vitamin D. Ruminococcus gnavus decreased with Vitamin D treatment in Cord. Supplementing with prenatal Vitamin D led to a rise in Ruminococcus gnavus.

3.3. *Association of Vitamin D with Microbial Composition and Diversity*

Nine out of 10 studies showed a significant association between Vitamin D and the gut microbiome (Table 2). There was some evidence that Vitamin D was associated with bacterial community composition. Four studies [36,37,40,43] determined that supplementation with Vitamin D_3 was associated with a shift in microbial composition. Two of these studies containing exclusively breast-fed infants showed a relation between Vitamin D supplementation and alpha and beta diversity. Vitamin D increased the microbial diversity and richness and showed a significant beta diversity [36,37,40,43]. On the contrary, a study in infants with eczema showed a significant low microbial diversity with high doses of Vitamin D but had similarities between the low-Vitamin D-supplemented group and healthy infants [40]. A positive association with the infant gut microbial community with Vitamin D status was observed in some of these studies.

Table 2. Microbial diversity of the selected studies on Vitamin D levels in infants.

Reference	Dosage	Study Time	Placebo or Comparator	Health Condition	Alpha Diversity and Beta Diversity of the Gut Microbiota
[34]	4000 IU vitamin D + prenatal vitamins or 400 IU vitamin D + prenatal vitamins	18 months	Different doses	Infants of asthmatic or allergic parent	No significant differences in the alpha diversity or beta diversity of the infant gut microbiota ($p > 0.05$)
[35]	2800 IU/day or 400 IU/day + placebo	3 months	Different doses with Placebo	Wheezing or asthmatic mothers	No significant differences were observed between the Vitamin D supplementation group and the control group for the bacterial diversity at 1 week, 1 month, or 1 year postpartum ($p = 0.955$, $p = 0.865$, $p = 0.971$).
[36]	400 IU of vitamin D/day or placebo	4 months	Placebo	Healthy	For the beta diversity, the gut bacteria composition differed significantly at the different time points (ANOSIM, R = 0.742, $p = 2.14 \times 10^{-5}$). The alpha diversity index strongly reduced in infants during the period from birth to term age while having no obvious change after 1 month
[37]	Vitamin D supplement vs. no Vitamin D supplement in past 24 h	24 h	Vitamin D doses or no doses	Healthy	Alpha diversity: lower Shannon index (p-value < 0.007) than those infants who were not given a Vitamin D. Beta diversity: p-value = 0.02
[38]	Maternal VDS: 0 µg or <10 µg/day or ≥10 µg/day. Infant VDS: classified as yes or no	6 months	Different doses or no dose	Healthy	N/A
[39]	50,000 IU/week	9 weeks	Before and after supplementation	Healthy	N/A
[40]	Vitamin D 400 IU/day or vitamin D 400 IU/day + vitamin D 200 (D-LOW) or 400 IU/day + vitamin D 400 IU/day (D-MED) or 400 IU/day+ 800 IU/day (D-HIGH)	1 month	Comparison between doses	Healthy infants and infant with Eczema	For alpha diversity, no significant difference in the richness in the groups. Diversity in the D–HIGH group was lower ($p < 0.01$). For beta diversity, there were similarities between eczema and D–MID group, as well as control and D-LOW group. Other groups differed significantly.
[41]	N/A	N/A	Vitamin D deficient and non-deficient	Healthy	No significant difference in alpha diversity beta diversity measure ($p < 0.01$)

Table 2. *Cont.*

Reference	Dosage	Study Time	Placebo or Comparator	Health Condition	Alpha Diversity and Beta Diversity of the Gut Microbiota
[42]	Maternal and infant VDS of >400 IU/day/ <400 IU/day/ 400 IU/day	3 months	Different doses or no dose	Healthy	N/A
[43]	N/A	N/A	Vitamin D levels	Higher risk for asthma	Higher prenatal 25(OH)D level was significantly associated with decreased richness ($p = 0.028$) and diversity ($p = 0.012$) of the gut microbiota at 1-month infants. Prenatal and cord 25(OH)D levels were significantly associated with 1-month microbiota composition.

3.4. *Vitamin D Effects on the Phyla*

Actinobacteria, Bacteroidetes, Firmicutes, and Proteobacteria were recorded as the most prevalent phyla in some of the studies [34,37]. All the studies showed alterations in the taxa of these phyla: Actinobacteria (Bifidobacterium spp and Corynebacterium), Bacteroidetes (Prevotellaceae, Prevotella, Prevotella copri, Alistipes, *Bacteroides* spp., Bacteroides massillensis, Bacteroides caceae, Bacteroides eggerthii, Bacteroides plebeius, Bacteroides vulgatus, and B. fragilis), Firmicutes (Lachnospiraceae, unclassified clostardilas, Lactococcus, Lactobacillus, Ruminococcus gnavus, C.difficile, Enterococcus, Acidaminococcales, Phascolarctobacterium, Phascolarctobacterium faecium, and Megamonas), Proteobacteria (Haemophilus, Bilophila, Sutterellaceae, Parasutterella, Parasutterella excrementihominis, and Acinetobacter) (Table 3).

Table 3. Alteration of bacterial taxa by Vitamin D in the selected studies.

Phylum	Class	Order	Family	Genus	Species
Actinobacteria	Actinomycetia	Mycobacteriales	Corynebacteriaceae	Corynebacterium	
Actinobacteria	Actinomycetia	Bifidobacteriales	Bifidobacteriaceae	*Bifidobacterium* spp.	
Bacteroidetes	Bacteroidia	Bacteroidales	Prevotellaceae	Prevotella	Prevotella dislens
Bacteroidetes	Bacteroidia	Bacteroidales	Prevotellaceae	Prevotella	Prevotella bivia
Bacteroidetes	Bacteroidia	Bacteroidales	Prevotellaceae	Prevotella	Prevotella copri
Bacteroidetes	Bacteroidia	Bacteroidales	Rikenellaceae	Alistipes	Alistipes finegoldii
Bacteroidetes	Bacteroidia	Bacteroidales	Rikenellaceae	Alistipes	Alistipes sp. AL1
Bacteroidetes	Bacteroidia	Bacteroidales	Rikenellaceae	Alistipes	Alistipes sp. N15 MGS-157
Bacteroidetes	Bacteroidia	Bacteroidales	Rikenellaceae	Alistipes	Alistipes marseille
Bacteroidetes	Bacteroidia	Bacteroidales	Rikenellaceae	Alistipes	Alistipes indistinctus
Bacteroidetes	Bacteroidia	Bacteroidales	Bacteroidaceae	Bacteroides	Bacteroides massillensis
Bacteroidetes	Bacteroidia	Bacteroidales	Bacteroidaceae	Bacteroides	Bacteroides caceae
Bacteroidetes	Bacteroidia	Bacteroidales	Bacteroidaceae	Bacteroides	Bacteroides eggerthii
Bacteroidetes	Bacteroidia	Bacteroidales	Bacteroidaceae	Bacteroides	Bacteroides plebeius
Bacteroidetes	Bacteroidia	Bacteroidales	Bacteroidaceae	Bacteroides	Bacteroides vulgatus
Firmicutes	Clostridia	Eubacteriales	Lachnospiraceae	Lachnobacterium	
Firmicutes	Clostridia	Clostridiales	Ruminococcaceae	Ruminococcus	Ruminococcus gnavus

Table 3. *Cont.*

Phylum	Class	Order	Family	Genus	Species
Firmicutes	Bacilli	Lactobacillales	Streptococcaceae	Lactococcus	
Firmicutes	Bacilli	Lactobacillales	Lactobacillaceae	Lactobacillus	
Firmicutes	Bacilli	Lactobacillales	Prcoccaceae	Streptococcus	
Firmicutes	Clostridia	Clostridiales	Unclassified Clostridiales		
Firmicutes	Bacilli	Lactobacillales	Enterococcaceae	Enterococcus	
Firmicutes	Clostridia	Eubacteriales	Peptostreptococcaceae	Clostridioides	C. difficile
Firmicutes	Negativicutes	Selenomonadales	Selenomonadaceae	Megamonas	
Firmicutes	Negativicutes	Acidaminococcales	Acidaminococcaceae	Phascolarctobacterium	Phascolarctobacterium faecium
Proteobacteria	Gammaproteobacteria	Pasteurellales	Pasteurellaceae	Haemophilus	
Proteobacteria	Gammaproteobacteria	Pseudomonadales	Moraxellaceae	Acinetobacter	
Proteobacteria	Deltaproteobacteria	Desulfovibrionales	Desulfovibrionaceae	Bilophila	
Proteobacteria	*Betaproteobacteria*	Burkholderiales	Sutterellaceae	Parasutterella	Parasutterella excrementihominis

taxa decreased with Vitamin D. taxa increased with Vitamin D. taxa that both increased and decreased with Vitamin D. taxa not significantly altered with Vitamin D. Green represents the taxa that were increased with Vitamin D levels, and red represents the taxa that were decreased with Vitamin D levels. Orange represents the taxa that were both increased and decreased and white represents the taxa that were unchanged.

There was a co-abundance of Firmicutes, Proteobacteria, Bacteriodetes, and Veillonella in children aged 3 to 6 months belonging to a parent with allergy or asthma [34]. Vitamin D levels increased the co-abundance of Firmicutes (Lachnospiraceae/Clostridiales) in these children and decreased Bacteroides. Firmicutes were significantly lower in Vitamin D-deficient children [42] but increased after supplementation in adolescent girls [40]. For breast-fed infants with 4 months supplementation of Vitamin D, they showed a significant increase in the abundance of Actinobacteria and a decrease in Bacteroidetes [37] in comparison with breast-fed infants without Vitamin D supplementation. Bacteroidetes and Lactobacillus were reduced by 72% and 24%, respectively, and Firmicutes and Enterococcus were increased [40] after high doses of Vitamin D supplementation. Proteobacteria was found to be the most abundant gut microbiome at birth. Infants whose mothers were supplemented with Vitamin D and breast-fed infants had decreased amounts of Proteobacteria, which is linked with obesity [43]. Children who were deficient in Vitamin D levels showed an elevated Bacteroidetes to-Firmicutes (B/F) ratio, which regulates gut microbiota homeostasis and also lowered the Bacteroides-to-Prevotella ratio, which is the main enterotype of the gut microbiome [41].

3.5. Alteration of the Other Taxa with Vitamin D

In the level of genus, Bifidobacterium was seen to be the most dominant genus present in faecal samples in three studies, regardless of any association with Vitamin D [36,37,40]. Bifidobacterium increased significantly after high doses of Vitamin D supplementation in 12-to-18-year-old females [40], and it inversely correlated with maternal multivitamin supplementation containing Vitamin D [39] and was higher in Vitamin D-supplemented breast-fed children.

Bacteroidetes and its species Bacteroides massillensis were increased in Vitamin D-deficient children [42], and some of its species (Bacteroides caceae, Bacteroides eggerthii, and Bacteroides plebeius) were lower in Vitamin D-deficient young children. Bacteroides were also lower in Vitamin D-supplemented infants. Prevotella was elevated in Vitamin D-deficient children, including species Prevotella dislens and Prevotella bivia [42]. Prvotella, its family Prevotellaceae, and species Prevotella copri were lowered, and the family Sutterellaceae, genus Parasutterella, and species Parasutterella excrementihominis increased with Vitamin D supplementation in children with eczema [40].

The genus Bilophila was found to be lowered in infants with prenatal Vitamin D supplementation [43]. *Alistipes* was determined to be significantly reduced in children with Vitamin D deficiency, including some species of the genus like, *Alistipes finegoldii*, *Alistipes* sp. AL1, and *Alistipes* sp. N15.MGS–157 [41], whereas some of the species of the genus increased in children with Vitamin D deficiency, for example, *Alistipes* marseille and *Alistipes* indistinctus [40]. Vitamin D supplementation had a significant reduction of Lactobacillus in adolescent girls [40] and increases in infants who were supplemented with Vitamin D [37]. Enterococcus Increased [40] and Streptococcus were found elevated with Vitamin D supplementation [37].

Maternal consumption of Vitamin D-fortified milk reduced the likelihood of C. difficile colonization in infants. C. difficile inversely correlated with prenatal multivitamin supplementation containing Vitamin D, and it was reduced in Vitamin D-supplemented infants who were breast-fed [39]. Prenatal maternal Vitamin D supplementation was not associated with infant C. difficile colonization in crude or adjusted models [43]. Order Acidaminococcales, genus Phascolarctobacterium, and species Phascolarctobacterium faecium increased with Vitamin D supplementation in infants with eczema [40].

Corynebacterium was higher with Vitamin D levels. Haemophilus was higher in prenatal supplementation and was found lower in infants who were supplemented with Vitamin D. Cord Blood Vitamin D levels were related with high Lachnobacterium but low Lactococcus [41]. Megamonas decreased in Vitamin D supplementation in infants [42], while Ruminococcus gnavus decreased with Vitamin D levels at the time of delivery and increased with prenatal Vitamin D levels [43].

4. Discussion

4.1. Pediatric Gut Microbiome and Immune System

In this review, we intended to study selected articles with changes in the gut microbiome in children with Vitamin D supplementation and different Vitamin D levels. The studies showed an association between Vitamin D and the intestinal microbiota in children, while the phyla Firmicutes, Bacteroidetes, Actinobacteria, and Proteobacteria were observed to be altered with Vitamin D. In diseases like NAFLD and ulcerative colitis, these phyla are considered important in the diseases' development [44,45]. A healthy pediatric gut microbiome is said to be dominated by Firmicutes, Bacteroidetes, Actinobacteria, and Proteobacteria [46,47]. An alteration of the taxa of these phyla are related with dysbiosis [48], which is related to diseases. A healthy gut microbiome has commensal and limited pathogenic microbes. The gut barrier plays an important part in maintaining a healthy microbiome by restricting pathogen entry. The immune cells can differentiate between commensal and pathogenic microbes with the help of the intestinal mucosal barrier, which reduces pathogenic reproduction [49]. Furthermore, the impermeability of the intestinal barrier helps reduce the adherence of the pathogens. The commensal bacteria, which have a neutral relationship with the host, can produce microbial metabolites such as bile acids, lactate, lipids, amino acids, vitamins like the Vitamin B complex, and SCFAs [3,50]. These gut microbiome metabolites are the major mediators for the immune system. For example, Parasutterella produces aromatic amino acid metabolites, which enhance gut barrier integrity [51]. SCFAs are also significant in keeping the integrity of the intestinal epithelial layer intact by stabilizing hypoxia-inducible factor 1, HIF-1. Higher partibility of the intestinal epithelium results from lower SCFAs, such as butyrate [52], restricting pathogen entry [49]. Thus, SCFAs and aromatic amino acids can help reduce dysbiosis and maintain intestinal homeostasis. This is supported by one of the selected studies where we saw SCFA-producing bacteria, acidaminococcus, and bacteroides vulgatus, while amino acid-producing bacteria Parasutterella were reduced in infants with eczema compared to healthy infants. The abundance of acidaminococcus, bacteroides vulgatus, and Parasutterella was elevated after Vitamin D supplementation in the infants in this study [40].

Overall, this implies that Vitamin D has an impact on the abundance of commensal bacteria. Commensal bacteria abundance is important to prevent pathogens, as these bacteria can compete with pathogenic bacteria for nutrients. Plus, Vitamin D receptors are present in the gut barrier and induce the production of antimicrobial peptides (AMPs) [53].

The Paneth cells in the GI tract release AMPs which kill pathogens by destroying their cell membranes. For instance, AMP α-defensins consist of human defensin 6, which does not allow the pathogenic bacteria to invade the gut, whereas β-defensins disrupt their cell membrane, and cathelicidin inhibit pathogen colonization [54]. AMPs modulate pH in order to prevent pathogens and reduce infectious diseases. Therefore, Vitamin D can alter the composition of the gut microbiome and the regulation of AMPs by Vitamin D, which leads them to modulate the immune system and reduce infections [55].

Vitamin D can also regulate the inflammatory response via SCFA production. Bacteria that produce SCFAs have an anti-inflammatory effect, which reduces inflammation. Butyrate raises transcription factor forkhead box P 3 (Foxp3) expression by T regulatory cells and histone acetylation [54], contributing further to the anti-inflammatory response. Additionally, it reduces nuclear factor-kappaB (NF-kB) [56], all of which reduces the production of pro-inflammatory cytokines. The increase of SCFA releasing bacteria by Vitamin D increases SCFA-influenced B-cell secretion of immunoglobin A (IgA). IgA plays an important part in maintaining gut barrier immune response as it reduces macrophages and dendritic cells and restricts chemotaxis in neutrophils [57]. Acetate-modulated IgA binds with specific microbes [58] as they take part in modulating inflammatory responses by elevating anti-inflammatory interleukin (IL), IL-10 and IL-18 as well as decreasing pro-inflammatory IL-1, TNF alpha, IL-6, IL-8, and IL-12 [57,59]. In the selected studies, Vitamin D reduced pro-inflammatory bacteria Bacteroidetes. Bacteroidetes and its genus Prevotella influence a pro-inflammatory response by releasing lipopolysaccharides, which activates macrophages and enhances inflammation. Prevotella was observed to be high in inflammatory disease, like eczema, and linked with declining health.

4.2. Vitamin D and Immune Response

Along with the modulation of the composition of the gut microbial flora, Vitamin D also regulates the human immune response towards microbiota. Vitamin D and its nuclear hormone receptor, VDR, can be expressed in the gut [34] and in immune cells like B and T lymphocytes [60]. The Vitamin D and VDR complex modulate the immune system by increasing some immune cells such as natural killer (NK), regulatory T cells (Treg), and IL-10, and decreasing some that include cells like neutrophil, IL-6, and IgE [61]. In one of the studies, the children with eczema had higher IL-6 and IgE levels, and after Vitamin D supplementation, both IgE and pro-inflammatory IL-6 were reduced to near-normal levels [40].

Vitamin D decreases the maturation of dendritic cells, increases macrophage M2, and reduces macrophage M1 phenotype. M1 activity inhibits cell proliferation and causes tissue damage, while M2 activity promotes cell proliferation and tissue repair. M1 macrophage releases TNF-a, IL-6, and cyclooxygenase-2 (COX2), which causes inflammation [62]. Vitamin D blocks the NF-kB pathway and COX2 transcription. This results in the reduction of reactive oxidative species. If not reduced, ROS can cause oxidative stress [63], DNA damage, and cell death.

The Vitamin D/VDR complex promotes T cell proliferation to form Treg cells and increase Th-2 pro-inflammatory cytokines, such as IL-10 cells. It also reduces Th-1 and Th-17 pro-inflammatory cytokines, such as IL-12 and IL-17. The increase of Th17 causes autoimmune diseases and allergies. As Vitamin D is responsible for increasing Treg cells and reducing cytokines IL-23 and IL-6, it reduces Th-17. It is also accompanied by the increase of Foxp3. Vitamin D decreases IFN-y, and that in turn reduces Th-1. When Vitamin D increases IL-4, it helps elevate Th2. Vitamin D also controls the B regulatory cells, decreases IgE release by B cells, and increases IL-10 [64].

4.3. The Role of Vitamin D in Pediatric Gut Microbiome and Diseases

As Vitamin D is linked with changing the gut microbiome and influencing the immune response via the gut microbiome, Vitamin D might also affect the immunological diseases relating to gut microbiome. Previously, studies correlated Vitamin D deficiency with increased autoimmune diseases. For example, *Megamonas* spp. increased the risk of childhood asthma, and decreased Lachnospira was found in asthmatic patients. These taxa were altered with increased Vitamin D levels. One of the studies showed a decrease of Megamonas, and another showed an increase of Lachnospira with Vitamin D supplementation [43]. In childhood asthma, an influx of pro-inflammatory IL-17, IL-3, and IL-4, as well as an increase of Th-17 cells and an imbalance of Th1/Th2 cells, are found responsible for the disease [65]. The pro-inflammatory cytokines and Th-17 decrease with elevated SCFA-releasing bacteria. Dysbiosis due to low levels of Vitamin D can cause less SCFA, which reduces dendritic cell maturation and, in turn, affects Th17. The reduction of Il-17 was also seen in allergic rhinitis patients with Vitamin D supplementation [65]. The reduction of IL-17, Il-4, and IgE were all linked with higher Vitamin D levels and an improvement of allergic symptoms. Allergic pediatric patients have elevated Bacteroides and Alistipes [66], both of which are linked with reduced SCFAs. In our selected studies, we found a decrease in Bacteroides and Alistipes with increasing Vitamin D.

Vitamin D deficiency was seen in Chinese children with Type 1 diabetes [67], and dysbiosis related to low Vitamin D levels was found in these children. This might be because butyrate reduction alters the tight junctions of the colon and causes leaky gut, which has been linked with diabetes. Lowering the abundance of Firmicutes can cause this imbalance of butyrate. In our search, Firmicutes phylum was observed to be increased with Vitamin D supplementation. Previous studies showed some Firmicutes were reduced in Cystic Fibrosis in children [68], and high amounts of some Firmicutes were related with obesity in children and C-reactive protein, which is an inflammation marker. In one of the studies, it was seen that the increasing Firmicutes and Bacteroidetes ratio was related with decreasing BMI [69]. Firmicutes and Bacteroidetes are said to be the two most significant phylum in the gut microbiome, to an extent that F/B ratio is considered as a determining factor of gut microbiome dysbiosis. An increase in the ratio can lead to IBD, and a decrease in its ratio can lead to obesity [70]. Similarly, we observed in one of the studies that the F/B ratio was lower in Vitamin D-deficient children. Relative to Firmicutes, Bacteroidetes was higher in deficient children, making a lower F/B ratio, which is known to maintain the balance of the microbiome [71]. The gut microbiota of 4–14-year-old children were profusely dominated by Prevotella. A higher Prevotella-to-Bacteroides (P/B) ratio was found in Vitamin D-deficient children, which has been linked with obesity [72]. Prevotella and Bacteroides are the main enterotypes that are usually found and help in classification of the intestinal microbiome. Hence, it indicates that Vitamin D supplementation can shift the enterotypes of the gut.

High Vitamin D levels are linked with reducing inflammatory disorders in children, like Crohn's disease [73]. A study suggests VDR can reduce inflammation in the ulcerative where hypoxia-inducible factor 1 is increased by blocking the NF-kB pathway [74]. Diseases like IBD, diarrhea, chronic inflammation, Type 1 diabetes, and infections are related to compromised gut epithelial layers [75]. The intestinal barrier integrity is upheld by the help of Vitamin D as it maintains which immune cells are being expressed in the tight junction, a layer that is significant in preventing bacterial invasion [76]. Basically, the Vitamin D/VDR complex suppresses apoptosis in the epithelium and restricts the inflammatory response to cause apoptosis in IBD patients. Vitamin D-influenced AMPs play critical roles in maintaining tolerance to gut microbiota as well, given that disruptions in tolerance to commensal microbiota and loss of barrier function play major roles in the pathogenesis of inflammatory bowel disease (IBD) [77]. The improvement of IBD symptoms after Vitamin D supplementation is consistent. NOD2, a nucleotide-binding oligomerization domain containing 2 from neutrophil-to-lymphocyte ratio (NLR) are linked with IBD. NOD2 can influence NF-kB as it binds with muramyl dipeptide and influences Transforming growth factor-β (TGF-β)-activated kinase 1 (TAK1). TAK1 is responsible for inititating NF-kB [78]. This results in autophagy. Vitamin D is associated with autophagy through the VD receptor (VDR). The loss of VDR impairs autophagy and enhances cell death through apoptosis. Dysbiosis in diseases like IBD is linked to NOD2 alleles, which is also related with Vitamin

D [12,78]. Vitamin D can regulate AMP production. It has been demonstrated that the level of expression of AMP can influence the relative composition of the gut microbiome in α-defensins deficient mice. So, Vitamin D levels can change the gut microbiome (Figure 2).

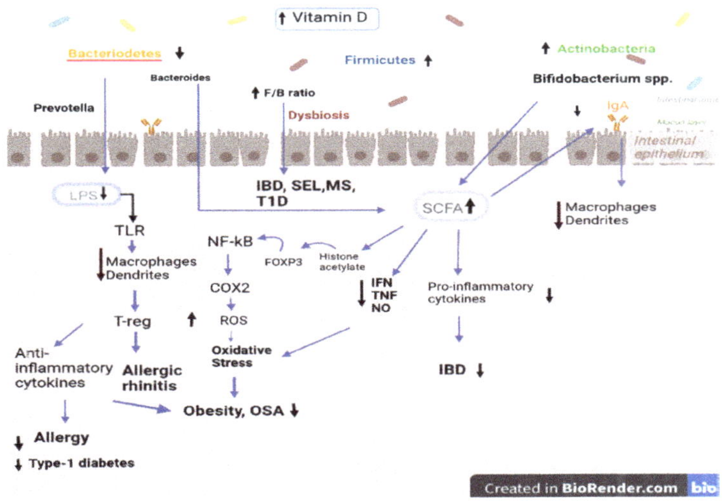

Figure 2. The role of Vitamin D in pediatric gut microbiome and diseases. Created with biorender.com (accessed on 13 January 2023).

4.4. Limitations

However, despite the evidence provided in this review that Vitamin D can influence the gut microbiome in children, there were some limitations to our study. Although we have done a thorough systematic search, we cannot exclude the possibility of not including papers outside the investigation protocol. Additionally, there was no statistical measurement performed. Even though there were not many methodological differences in the studies, this can be considered for further analysis.

5. Conclusions

The studies included in this review have shown evidence that the composition of the children's intestinal microbiome could be altered by Vitamin D levels. These present data suggesting an indirect influence of Vitamin D on the children's gut microbiome. The pediatric gut microbiome was altered in Vitamin D deficiency and supplementation during the prenatal and infancy period, as well as with older children. Therefore, Vitamin D plays an important role in altering the gut microbiome in children. Vitamin D supplementation has been determined to improve gut microbiome and the diversity in children, which can have a positive impact on their overall health and immune system function. Furthermore, when evaluating Vitamin D as a potential treatment for immunological illnesses in children associated with gut microbiota, such as allergies, asthma, and eczema, the evaluation of publications in this systematic review can be used as a source of information. No systematic review, to our knowledge, has reported and evaluated this link in children. However, more research is needed to fully understand the mechanisms behind the relationship between Vitamin D and gut microbiota and how this relationship can be utilized for therapeutic interventions. These data could potentially implicate future clinical studies to observe the direct effect of Vitamin D supplementation towards reducing the clinical symptoms of immunological diseases. Nonetheless, ensuring that children have adequate Vitamin D levels through diet or supplementation may be an important strategy for maintaining a healthy gut microbiome and promoting overall health in children.

Author Contributions: Conceptualization, N.A.S.I. and A.A.; methodology, A.T.; validation, N.A.S.I., A.A. and F.D.Z.; formal analysis, A.T. and N.A.S.I. data curation, A.T.; writing—original draft preparation, A.T.; writing—review and editing, N.A.S.I., A.A. and F.D.Z.; visualization, A.T.; supervision, N.A.S.I., A.A. and F.D.Z.; project administration, N.A.S.I.; funding acquisition, N.A.S.I. All authors have read and agreed to the published version of the manuscript.

Funding: This research received no external funding.

Institutional Review Board Statement: Not applicable.

Informed Consent Statement: Not applicable.

Data Availability Statement: Not applicable.

Acknowledgments: We are grateful to the Faculty of Medicine, Universiti Kebangsaan Malaysia for sponsoring this publication.

Conflicts of Interest: The authors declare no conflict of interest.

Appendix A

Table A1. Search strategy for PubMed/SCOPUS/WOS/MEDLINE/OVID.

No.	Search Strategy
#1	Vitamin D AND microbiome
#2	Vitamin D AND microbiome AND immun *
#3	Vitamin D AND microbiome AND infection
#4	Vitamin D AND microbiome AND inflammat *
#5	Vitamin D AND microbiome AND profiling
#6	Vitamin D AND microbiome AND transcript *
#7	Vitamin D AND microbiome AND pediatric
#8	Vitamin D AND microbiome AND immune AND infection AND inflammation AND profiling AND transcript * AND pediatric

(*) Used as a truncation symbol

Table A2. Initial search strategy overview.

Databases	PubMed, SCOPUS, WOS, MEDLINE/OVID
Date of search	from January 1996 to January 2023
Study type	Books and documents were excluded
Limit	Conference abstracts were excluded
Other databases	Google Scholar was used with the same keywords as the main databases

References

1. Fan, Y.; Pedersen, O. Gut Microbiota in Human Metabolic Health and Disease. *Nat. Rev. Microbiol.* **2020**, *19*, 55–71. [CrossRef] [PubMed]
2. Yukgehnaish, K.; Kumar, P.; Sivachandran, P.; Marimuthu, K.; Arshad, A.; Paray, B.A.; Arockiaraj, J. Gut Microbiota Metagenomics in Aquaculture: Factors Influencing Gut Microbiome and Its Physiological Role in Fish. *Rev. Aquac.* **2020**, *12*, 1903–1927. [CrossRef]
3. Xiao, H.; Kang, S. The Role of the Gut Microbiome in Energy Balance With a Focus on the Gut-Adipose Tissue Axis. *Front. Genet.* **2020**, *11*, 297. [CrossRef]
4. Sanna, S.; van Zuydam, N.R.; Mahajan, A.; Kurilshikov, A.; Vich Vila, A.; Võsa, U.; Mujagic, Z.; Masclee, A.A.M.; Jonkers, D.M.A.E.; Oosting, M.; et al. Causal Relationships among the Gut Microbiome, Short-Chain Fatty Acids and Metabolic Diseases. *Nat. Genet.* **2019**, *51*, 600–605. [CrossRef]
5. Das, P.; Babaei, P.; Nielsen, J. Metagenomic analysis of microbe-mediated vitamin metabolism in the human gut microbiome. *BMC Genom.* **2019**, *20*, 208. [CrossRef]
6. Bosco, N.; Noti, M. The aging gut microbiome and its impact on host immunity. *Genes. Immun.* **2021**, *22*, 289–303. [CrossRef] [PubMed]
7. Ronan, V.; Yeasin, R.; Claud, E.C. Childhood Development and the Microbiome-The Intestinal Microbiota in Maintenance of Health and Development of Disease During Childhood Development. *Gastroenterology* **2021**, *160*, 495–506. [CrossRef] [PubMed]
8. Kapourchali, F.R.; Cresci, G.A.M. Early-Life Gut Microbiome-The Importance of Maternal and Infant Factors in Its Establishment. *Nutr. Clin. Pr.* **2020**, *35*, 386–405. [CrossRef]
9. Saeed, N.K.; Al-Beltagi, M.; Bediwy, A.S.; El-Sawaf, Y.; Toema, O. Gut microbiota in various childhood disorders: Implication and indications. *World J. Gastroenterol.* **2022**, *28*, 1875–1901. [CrossRef]

10. Chiu, C.Y.; Chan, Y.L.; Tsai, M.H.; Wang, C.J.; Chiang, M.H.; Chiu, C.C. Gut microbial dysbiosis is associated with allergen-specific IgE responses in young children with airway allergies. *World Allergy Organ. J.* **2019**, *12*, 100021. [CrossRef] [PubMed]
11. Harbison, J.E.; Roth-Schulze, A.J.; Giles, L.C.; Tran, C.D.; Ngui, K.M.; Penno, M.A.; Thomson, R.L.; Wentworth, J.M.; Colman, P.G.; Craig, M.E.; et al. Gut Microbiome Dysbiosis and Increased Intestinal Permeability in Children with Islet Autoimmunity and Type 1 Diabetes: A Prospective Cohort Study. *Pediatr. Diabetes* **2019**, *20*, 574–583. [CrossRef]
12. Fitzgerald, R.S.; Sanderson, I.R.; Claesson, M.J. Paediatric Inflammatory Bowel Disease and its Relationship with the Microbiome. *Microb. Ecol.* **2021**, *82*, 833–844. [CrossRef] [PubMed]
13. Tremlett, H.; Zhu, F.; Arnold, D.; Bar-Or, A.; Bernstein, C.N.; Bonner, C.; Forbes, J.D.; Graham, M.; Hart, J.; Knox, N.C.; et al. The Gut Microbiota in Pediatric Multiple Sclerosis and Demyelinating Syndromes. *Ann. Clin. Transl. Neurol.* **2021**, *8*, 2252–2269. [CrossRef]
14. Hufnagl, K.; Pali-Schöll, I.; Roth-Walter, F.; Jensen-Jarolim, E. Dysbiosis of the gut and lung microbiome has a role in asthma. *Semin. Immunopathol.* **2020**, *42*, 75–93. [CrossRef] [PubMed]
15. Zhong, H.; Penders, J.; Shi, Z.; Ren, H.; Cai, K.; Fang, C.; Ding, Q.; Thijs, C.; Blaak, E.E.; Stehouwer, C.D.A.; et al. Impact of Early Events and Lifestyle on the Gut Microbiota and Metabolic Phenotypes in Young School-Age Children. *Microbiome* **2019**, *7*, 2. [CrossRef]
16. McDonnell, L.; Gilkes, A.; Ashworth, M.; Rowland, V.; Harries, T.H.; Armstrong, D.; White, P. Association between Antibiotics and Gut Microbiome Dysbiosis in Children: Systematic Review and Meta-Analysis. *Gut Microbes* **2021**, *13*, 1870402. [CrossRef]
17. Galazzo, G.; van Best, N.; Bervoets, L.; Dapaah, I.O.; Savelkoul, P.H.; Hornef, M.W.; Lau, S.; Hamelmann, E.; Penders, J.; Hutton, E.K.; et al. Development of the Microbiota and Associations with Birth Mode, Diet, and Atopic Disorders in a Longitudinal Analysis of Stool Samples, Collected from Infancy through Early Childhood. *Gastroenterology* **2020**, *158*, 1584–1596. [CrossRef] [PubMed]
18. Fehr, K.; Moossavi, S.; Sbihi, H.; Boutin, R.C.T.; Bode, L.; Robertson, B.; Yonemitsu, C.; Field, C.J.; Becker, A.B.; Mandhane, P.J.; et al. Breastmilk Feeding Practices Are Associated with the Co-Occurrence of Bacteria in Mothers' Milk and the Infant Gut: The CHILD Cohort Study. *Cell Host Microbe* **2020**, *28*, 285–297.e4. [CrossRef]
19. Clark, A.; Mach, N. Role of vitamin D in the hygiene hypothesis: The interplay between vitamin D, vitamin D receptors, gut microbiota, and immune response. *Front. Immunol.* **2016**, *7*, 627. [CrossRef]
20. Singh, P.; Rawat, A.; Alwakeel, M.; Sharif, E.; Al Khodor, S. The potential role of vitamin D supplementation as a gut microbiota modifier in healthy individuals. *Sci. Rep.* **2020**, *10*, 21641. [CrossRef]
21. Zhou, H.; Zhao, X.; Sun, L.; Liu, Y.; Lv, Y.; Gang, X.; Wang, G. Gut Microbiota Profile in Patients with Type 1 Diabetes Based on 16S RRNA Gene Sequencing: A Systematic Review. *Dis. Mrk.* **2020**, *2020*, 1–11. [CrossRef]
22. Almonacid, D.E.; Kraal, L.; Ossandon, F.J.; Budovskaya, Y.V.; Cardenas, J.P.; Bik, E.M.; Goddard, A.D.; Richman, J.; Apte, Z.S. Correction: 16S RRNA Gene Sequencing and Healthy Reference Ranges for 28 Clinically Relevant Microbial Taxa from the Human Gut Microbiome. *PLoS ONE* **2019**, *14*, e0212474. [CrossRef] [PubMed]
23. Durazzi, F.; Sala, C.; Castellani, G.; Manfreda, G.; Remondini, D.; De Cesare, A. Comparison between 16S rRNA and shotgun sequencing data for the taxonomic characterization of the gut microbiota. *Sci. Rep.* **2021**, *11*, 3030. [CrossRef]
24. Bellerba, F.; Muzio, V.; Gnagnarella, P.; Facciotti, F.; Chiocca, S.; Bossi, P.; Cortinovis, D.; Chiaradonna, F.; Serrano, D.; Raimondi, S.; et al. The association between Vitamin D and gut microbiota: A systematic review of human studies. *Nutrients* **2021**, *13*, 3378. [CrossRef] [PubMed]
25. Cyprian, F.; Lefkou, E.; Varoudi, K.; Girardi, G. Immunomodulatory Effects of Vitamin D in Pregnancy and Beyond. *Front. Immunol.* **2019**, *10*, 2739. [CrossRef] [PubMed]
26. Kalmarzi, R.N.; Ahmadi, S.; Rahehagh, R.; Fathallahpour, A.; Khalafi, B.; Kashefi, H.; Roshani, D.; Zaryan, R.N.; Mohamadi, S.; Kooti, W. The Effect of Vitamin D Supplementation on Clinical Outcomes of Asthmatic Children with Vitamin D Insufficiency. *Endocr. Metab. Immune Disord.-Drug. Targets* **2020**, *20*, 149–155. [CrossRef]
27. Bhardwaj, B.; Singh, J. Efficacy of Vitamin D Supplementation in Allergic Rhinitis. *Indian. J. Otolaryngol. Head. Neck Surg.* **2021**, *73*, 152–159. [CrossRef]
28. Guo, Y.; Zhang, T.; Wang, Y.; Liu, R.; Chang, M.; Wang, X. Effects of oral vitamin D supplementation on inflammatory bowel disease: A systematic review and meta-analysis. *Food Funct.* **2021**, *12*, 7588–7606. [CrossRef]
29. Fakhoury, H.M.A.; Kvietys, P.R.; AlKattan, W.; Anouti, F.A.; Elahi, M.A.; Karras, S.N.; Grant, W.B. Vitamin D and Intestinal Homeostasis: Barrier, Microbiota, and Immune Modulation. *J. Steroid Biochem. Mol. Biol.* **2020**, *200*, 105663. [CrossRef]
30. Wu, Z.; Liu, D.; Deng, F. The Role of Vitamin D in Immune System and Inflammatory Bowel Disease. *J. Inflamm. Res.* **2022**, *15*, 3167–3185. [CrossRef]
31. Tang, W.; Rong, Y.; Zhang, H.; Zhan, Z.; Yuan, L.; Ning, Y.; Lin, W. The Correlation between a Th1/Th2 Cytokines Imbalance and Vitamin D Level in Patients with Early Chronic Obstructive Pulmonary Disease (COPD), Based on Screening Results. *Front. Physiol.* **2023**, *14*, 1032786. [CrossRef]
32. Luthold, R.V.; Fernandes, G.R.; Franco-de-Moraes, A.C.; Folchetti, L.G.D.; Ferreira, S.R.G. Gut microbiota interactions with the immunomodulatory role of vitamin D in normal individuals. *Metabolism* **2017**, *69*, 76–86. [CrossRef] [PubMed]
33. Zeng, Y.; Luo, M.; Pan, L.; Chen, Y.; Guo, S.; Luo, D.; Zhu, L.; Liu, Y.; Pan, L.; Xu, S.; et al. Vitamin D Signaling Maintains Intestinal Innate Immunity and Gut Microbiota: Potential Intervention for Metabolic Syndrome and NAFLD. *Am. J. Physiol.-Gastrointest. Liver Physiol.* **2020**, *318*, G542–G553. [CrossRef]

34. Sordillo, J.E.; Zhou, Y.; McGeachie, M.J.; Ziniti, J.; Lange, N.; Laranjo, N.; Savage, J.R.; Carey, V.; O'Connor, G.; Sandel, M.; et al. Factors Influencing the Infant Gut Microbiome at Age 3–6 Months: Findings from the Ethnically Diverse Vitamin D Antenatal Asthma Reduction Trial (VDAART). *J. Allergy Clin. Immunol.* **2017**, *139*, 482–491.e14. [CrossRef]
35. Hjelmsø, M.H.; Shah, S.A.; Thorsen, J.; Rasmussen, M.; Vestergaard, G.; Mortensen, M.S.; Brejnrod, A.; Brix, S.; Chawes, B.; Bønnelykke, K.; et al. Prenatal Dietary Supplements Influence the Infant Airway Microbiota in a Randomized Factorial Clinical Trial. *Nat. Commun.* **2020**, *11*, 426. [CrossRef] [PubMed]
36. Lei, W.-T.; Huang, K.-Y.; Jhong, J.-H.; Chen, C.-H.; Weng, S.-L. Metagenomic Analysis of the Gut Microbiome Composition Associated with Vitamin D Supplementation in Taiwanese Infants. *Sci. Rep.* **2021**, *11*, 2856. [CrossRef] [PubMed]
37. Ma, T.; Bu, S.; Paneth, N.; Kerver, J.M.; Comstock, S.S. Vitamin D Supplementation in Exclusively Breastfed Infants Is Associated with Alterations in the Fecal Microbiome. *Nutrients* **2022**, *14*, 202. [CrossRef]
38. Talsness, C.E.; Penders, J.; Jansen, E.H.J.M.; Damoiseaux, J.; Thijs, C.; Mommers, M. Influence of Vitamin D on Key Bacterial Taxa in Infant Microbiota in the KOALA Birth Cohort Study. *PLoS ONE* **2017**, *12*, e0188011. [CrossRef]
39. Tabatabaeizadeh, S.-A.; Fazeli, M.; Meshkat, Z.; Khodashenas, E.; Esmaeili, H.; Mazloum, S.; Ferns, G.A.; Abdizadeh, M.F.; Ghayour-Mobarhan, M. The Effects of High Doses of Vitamin D on the Composition of the Gut Microbiome of Adolescent Girls. *Clin. Nutr. ESPEN* **2020**, *35*, 103–108. [CrossRef]
40. Liu, Y.; Yang, M.; Ran, Z.; Wang, J.; Ma, W.; Sheng, Q. Effect of Different Doses of Vitamin D on the Intestinal Flora of Babies with Eczema: An Experimental Study. *Life* **2022**, *12*, 1409. [CrossRef]
41. Singh, P.; Rawat, A.; Saadaoui, M.; Elhag, D.; Tomei, S.; Elanbari, M.; Akobeng, A.K.; Mustafa, A.; Abdelgadir, I.; Udassi, S.; et al. Tipping the Balance: Vitamin D Inadequacy in Children Impacts the Major Gut Bacterial Phyla. *Biomedicines* **2022**, *10*, 278. [CrossRef]
42. Drall, K.M.; Field, C.J.; Haqq, A.M.; de Souza, R.J.; Tun, H.M.; Morales-Lizcano, N.P.; Konya, T.B.; Guttman, D.S.; Azad, M.B.; Becker, A.B.; et al. Vitamin D Supplementation in Pregnancy and Early Infancy in Relation to Gut Microbiota Composition and C. *Difficile* Colonization: Implications for Viral Respiratory Infections. *Gut Microbes* **2020**, *12*, 1799734. [CrossRef] [PubMed]
43. Kassem, Z.; Sitarik, A.; Levin, A.M.; Lynch, S.V.; Havstad, S.; Fujimura, K.; Kozyrskyj, A.; Ownby, D.R.; Johnson, C.C.; Yong, G.J.M.; et al. Maternal and Cord Blood Vitamin D Level and the Infant Gut Microbiota in a Birth Cohort Study. *Matern. Health Neonatol. Perinatol.* **2020**, *6*, 5. [CrossRef]
44. Jasirwan, C.O.M.; Muradi, A.; Hasan, I.; Simadibrata, M.; Rinaldi, I. Correlation of Gut Firmicutes/Bacteroidetes Ratio with Fibrosis and Steatosis Stratified by Body Mass Index in Patients with Non-Alcoholic Fatty Liver Disease. *Biosci. Microbiota Food Health* **2020**, *40*, 50–58. [CrossRef]
45. Zakerska-Banaszak, O.; Tomczak, H.; Gabryel, M.; Baturo, A.; Wolko, L.; Michalak, M.; Malinska, N.; Mankowska-Wierzbicka, D.; Eder, P.; Dobrowolska, A.; et al. Dysbiosis of Gut Microbiota in Polish Patients with Ulcerative Colitis: A Pilot Study. *Sci. Rep.* **2021**, *11*, 2166. [CrossRef]
46. Deering, K.E.; Devine, A.; O'sullivan, T.A.; Lo, J.; Boyce, M.C.; Christophersen, C.T. Characterizing the Composition of the Pediatric Gut Microbiome: A Systematic Review. *Nutrients* **2019**, *12*, 16. [CrossRef] [PubMed]
47. Das, B.; Nair, G.B. Homeostasis and dysbiosis of the gut microbiome in health and disease. *J. Biosci.* **2019**, *44*, 1–8. [CrossRef]
48. Antonini, M.; Conte, M.L.; Sorini, C.; Falcone, M. How the interplay between the commensal microbiota, gut barrier integrity and mucosal immunity regulates brain autoimmunity. *Front. Immunol.* **2019**, *10*, 1937. [CrossRef]
49. Rinninella, E.; Raoul, P.; Cintoni, M.; Franceschi, F.; Miggiano, G.; Gasbarrini, A.; Mele, M. What Is the Healthy Gut Microbiota Composition? A Changing Ecosystem across Age, Environment, Diet, and Diseases. *Microorganisms* **2019**, *7*, 14. [CrossRef]
50. Eslami, M.; Bahar, A.; Hemati, M.; Rasouli Nejad, Z.; Mehranfar, F.; Karami, S.; Kobyliak, N.M.; Yousefi, B. Dietary Pattern, Colonic Microbiota and Immunometabolism Interaction: New Frontiers for Diabetes Mellitus and Related Disorders. *Diabetic Medicine* **2020**, *38*, 255. [CrossRef]
51. Martinez-Guryn, K.; Leone, V.; Chang, E.B. Regional Diversity of the Gastrointestinal Microbiome. *Cell Host Microbe* **2019**, *26*(3), 314–324. [CrossRef] [PubMed]
52. Wang, G.; Hu, Y.-X.; He, M.-Y.; Xie, Y.-H.; Su, W.; Long, D.; Zhao, R.; Wang, J.; Dai, C.; Li, H.; et al. Gut-Lung Dysbiosis Accompanied by Diabetes Mellitus Leads to Pulmonary Fibrotic Change through the NF-KB Signaling Pathway. *Am. J. Pathol.* **2021**, *191*, 838–856. [CrossRef] [PubMed]
53. White, J.H. Emerging Roles of Vitamin D-Induced Antimicrobial Peptides in Antiviral Innate Immunity. *Nutrients* **2022**, *14*, 284. [CrossRef] [PubMed]
54. Gubatan, J.; Rogalla, S.; Holman, D.R.; Puntasecca, C.J.; Polevoi, D.; Rubin, S.J.S. Antimicrobial peptides and the gut microbiome in inflammatory bowel disease. *World J. Gastroenterol.* **2021**, *27*, 506–520. [CrossRef] [PubMed]
55. Zhou, H.; Sun, L.; Zhang, S.; Zhao, X.; Gang, X.; Wang, G. Evaluating the Causal Role of Gut Microbiota in Type 1 Diabetes and Its Possible Pathogenic Mechanisms. *Front. Endocrinol.* **2020**, *11*, 1–12. [CrossRef]
56. Yao, Y.; Cai, X.; Fei, W.; Ye, Y.; Zhao, M.; Zheng, C. The role of short-chain fatty acids in immunity, inflammation and metabolism. *Crit. Rev. Food Sci. Nutr.* **2020**, *62*, 3803–3816. [CrossRef]
57. Takeuchi, T.; Miyauchi, E.; Kanaya, T.; Kato, T.; Nakanishi, Y.; Watanabe, T.; Kitami, T.; Taida, T.; Sasaki, T.; Negishi, H.; et al. Acetate Differentially Regulates IgA Reactivity to Commensal Bacteria. *Nature* **2021**, *595*, 560–564. [CrossRef]
58. Danne, C.; Sokol, H. Butyrate, a new microbiota-dependent player in CD8+ T cells immunity and cancer therapy? *Cell Rep. Med.* **2021**, *2*, 1032–1043. [CrossRef]

59. Zhang, Y.; Garrett, S.; Carroll, R.E.; Xia, Y.; Sun, J. Vitamin D receptor upregulates tight junction protein claudin-5 against colitis-associated tumorigenesis. *Mucosal Immunol.* **2022**, *15*, 683–697. [CrossRef]
60. Arora, J.; Wang, J.; Weaver, V.; Zhang, Y.; Cantorna, M.T. Novel Insight into the Role of the Vitamin D Receptor in the Development and Function of the Immune System. *J. Steroid Biochem. Mol. Biol.* **2022**, *219*, 106084. [CrossRef] [PubMed]
61. Zhang, X.; Zhao, Y.; Zhu, X.-D.; Guo, Y.; Yang, Y.; Jiang, Y.; Liu, B.-C. Active Vitamin D Regulates Macrophage M1/M2 Phenotypes via the STAT-1-TREM-1 Pathway in Diabetic Nephropathy. *J. Cell. Physiol.* **2019**, *234*, 6917–6926. [CrossRef]
62. Ricca, C.; Aillon, A.; Bergandi, L.; Alotto, D.; Castagnoli, C.; Silvagno, F. Vitamin D Receptor Is Necessary for Mitochondrial Function and Cell Health. *Int. J. Mol. Sci.* **2018**, *19*, 1672. [CrossRef] [PubMed]
63. Fantini, C.; Corinaldesi, C.; Lenzi, A.; Migliaccio, S.; Crescioli, C. Vitamin D as a Shield against Aging. *Int. J. Mol. Sci.* **2023**, *24*, 4546. [CrossRef] [PubMed]
64. Agarwal, S.; Singh, S.N.; Kumar, R.; Sehra, R. Vitamin D: A Modulator of Allergic Rhinitis. *Indian J. Otolaryngol. Head Neck Surg.* **2019**, *71*, 2225–2230. [CrossRef] [PubMed]
65. Kourosh, A.; Luna, R.A.; Balderas, M.; Nance, C.; Anagnostou, A.; Devaraj, S.; Davis, C.M. Fecal Microbiome Signatures Are Different in Food-Allergic Children Compared to Siblings and Healthy Children. *Pediatr. Allergy Immunol.* **2018**, *29*, 545–554. [CrossRef]
66. Liu, X.; Cheng, Y.-W.; Shao, L.; Sun, S.-H.; Wu, J.; Song, Q.-H.; Zou, H.-S.; Ling, Z.-X. Gut Microbiota Dysbiosis in Chinese Children with Type 1 Diabetes Mellitus: An Observational Study. *World J. Gastroenterol.* **2021**, *27*, 2394–2414. [CrossRef]
67. de Freitas, M.B.; Moreira, E.A.M.; Tomio, C.; Moreno, Y.M.F.; Daltoe, F.P.; Barbosa, E.; Ludwig Neto, N.; Buccigrossi, V.; Guarino, A. Altered Intestinal Microbiota Composition, Antibiotic Therapy and Intestinal Inflammation in Children and Adolescents with Cystic Fibrosis. *PLoS ONE* **2018**, *13*, e0198457. [CrossRef]
68. Metzger, M.I.; Graeber, S.Y.; Stahl, M.; Sommerburg, O.; Mall, M.A.; Dalpke, A.H.; Boutin, S. A Volatile and Dynamic Longitudinal Microbiome Is Associated with Less Reduction in Lung Function in Adolescents with Cystic Fibrosis. *Front. Cell. Infect. Microbiol.* **2021**, *11*, 2235–2988. [CrossRef] [PubMed]
69. Grigor'eva, I.N. Gallstone Disease, Obesity and the Firmicutes/Bacteroidetes Ratio as a Possible Biomarker of Gut Dysbiosis. *J. Pers. Med.* **2020**, *11*, 13. [CrossRef]
70. Stojanov, S.; Berlec, A.; Štrukelj, B. The Influence of Probiotics on the Firmicutes/Bacteroidetes Ratio in the Treatment of Obesity and Inflammatory Bowel Disease. *Microorganisms* **2020**, *8*, 1715. [CrossRef]
71. Dong, T.S.; Guan, M.; Mayer, E.A.; Stains, J.; Liu, C.; Vora, P.; Jacobs, J.P.; Lagishetty, V.; Chang, L.; Barry, R.L.; et al. Obesity Is Associated with a Distinct Brain-Gut Microbiome Signature That Connects Prevotella and Bacteroides to the Brain's Reward Center. *Gut Microbes* **2022**, *14*, 1–20. [CrossRef] [PubMed]
72. White, J.H. Vitamin D deficiency and the pathogenesis of Crohn's disease. *J. Steroid Biochem. Mol. Biol.* **2018**, *175*, 23–28. [CrossRef]
73. Xue, G.; Gao, R.; Liu, Z.; Xu, N.; Cao, Y.; Zhao, B.; Du, J. Vitamin D/VDR Signaling Inhibits Colitis by Suppressing HIF-1α Activation in Colonic Epithelial Cells. *Am. J. Physiol.-Gastrointest. Liver Physiol.* **2021**, *320*, G837–G846. [CrossRef] [PubMed]
74. Mailhot, G.; White, J.H. Vitamin D and Immunity in Infants and Children. *Nutrients* **2020**, *12*, 1233. [CrossRef]
75. Yang, Y.; Cui, X.; Li, J.; Wang, H.; Li, Y.; Chen, Y.; Zhang, H. Clinical Evaluation of Vitamin D Status and Its Relationship with Disease Activity and Changes of Intestinal Immune Function in Patients with Crohn's Disease in the Chinese Population. *Scand. J. Gastroenterol.* **2021**, *56*, 20–29. [CrossRef]
76. Battistini, C.; Ballan, R.; Herkenhoff, M.E.; Saad, S.M.I.; Sun, J. Vitamin D Modulates Intestinal Microbiota in Inflammatory Bowel Diseases. *Int. J. Mol. Sci.* **2020**, *22*, 362. [CrossRef] [PubMed]
77. Martens, P.-J.; Centelles-Lodeiro, J.; Ellis, D.; Cook, D.P.; Sassi, G.; Verlinden, L.; Verstuyf, A.; Raes, J.; Mathieu, C.; Gysemans, C. High Serum Vitamin D Concentrations, Induced via Diet, Trigger Immune and Intestinal Microbiota Alterations Leading to Type 1 Diabetes Protection in NOD Mice. *Front. Immunol.* **2022**, *13*, 1664–3224. [CrossRef]
78. Tabatabaeizadeh, S.-A.; Tafazoli, N.; Ferns, G.A.; Avan, A.; Ghayour-Mobarhan, M. Vitamin D, the gut microbiome and inflammatory bowel disease. *J. Res. Med. Sci.* **2018**, *23*, 75. [CrossRef]

biomedicines

MDPI

Article

Association between Organochlorine Pesticides and Vitamin D in Female Subjects

Edwina Brennan [1,*], Alexandra E. Butler [1], Manjula Nandakumar [1], Daniel S. Drage [2,3], Thozhukat Sathyapalan [4] and Stephen L. Atkin [1]

1 School of Medicine, Royal College of Surgeons in Ireland-Medical University of Bahrain, Busaiteen 15503, Bahrain; abutler@rcsi.com (A.E.B.); mnandakumar@rcsi.com (M.N.); satkin@rcsi.com (S.L.A.)
2 School of Geography, Earth and Environmental Sciences, University of Birmingham, Edgbaston, Birmingham B15 2TT, UK; d.s.drage@bham.ac.uk
3 Queensland Alliance for Environmental Health Sciences, The University of Queensland, 39 Kessels Road, Brisbane, QLD 4108, Australia
4 Hull York Medical School, University of Hull, Hull HU6 7RX, UK; thozhukat.sathyapalan@hyms.ac.uk
* Correspondence: ebrennan@rcsi.com

Abstract: In human population studies, organochlorine pesticides (OCPs) have been linked to vitamin D deficiency. Therefore, this study examined the association between OCPs, vitamin D_3 (cholecalciferol, 25(OH)D_3), and the active metabolite 1,25-dihydrovitamin D_3 (1,25$(OH)_2D_3$) in a cohort of non-obese women. The serum samples of 58 female participants (age—31.9 ± 4.6 years; body mass index (BMI)—25.7 ± 3.7 kg/m^2) were screened for 10 indicator OCPs. 25(OH)D_3 and 1,25$(OH)_2D_3$ levels were determined via isotope dilution liquid chromatography tandem mass spectrometry. In this cohort, the 25(OH)D_3 and 1,25$(OH)_2D_3$ levels were 22.9 ± 11.2 ng/mL and 0.05 ± 0.02 ng/mL, respectively, with 28 participants classified as 25(OH)D_3-deficient (<50 nmol/L). In the study cohort, no correlations were found between individual or total OCPs (ΣOCPs) and 25(OH)D_3. p,p′-dichlorodiphenyldichloroethylene (DDE) and ΣOCPs correlated positively with 1,25$(OH)_2D_3$, with the latter being negatively correlated with estimated glomerular filtration rate (eGFR). In women with sufficient 25(OH)D_3 levels, p,p′-dichlorodiphenyltrichloroethan (DDT) was positively correlated with 1,25$(OH)_2D_3$, whilst in the deficient group, hexachlorobenzene (HCB) and p,p′-(DDE) were positively correlated with 1,25$(OH)_2D_3$, β-Hexachlorocyclohexane (HCH) was positively correlated with 25(OH)D_3, and none of the OCPs were associated with measures of renal function. Overall, OCPs and ΣOCPs were not associated with 25(OH)D_3, suggesting that they are unrelated to vitamin D deficiency, but p,p′-DDE and ΣOCPs correlated positively with active 1,25$(OH)_2D_3$, while ΣOCPs correlated negatively with eGFR, suggesting a possible renal effect. Analysis of vitamin D deficiency revealed an association between β-HCH and 25(OH)D_3, and between HCB and p,p′-DDE and 1,25$(OH)_2D_3$, suggesting that OCP effects may be enhanced in cases of vitamin D deficiency.

Keywords: Organochlorine pesticides (OCPs); organic pollutants; vitamin D; cholecalciferol

1. Introduction

Organochlorine pesticides (OCPs) are a group of chlorinated hydrocarbons that were once used extensively in agriculture and mosquito control and have since been termed persistent organic pollutants (POPs) due to their high toxicity, slow degradation, and bioaccumulation [1]. OCPs are highly persistent in the environment and can accumulate in the fatty tissues of animals and humans [2,3]. They can also be transported over great distances in the air and water, leading to the contamination of soil and water sources far from the OCPs' original application site [4]. Due to the potential health risks associated with OCPs, many countries have banned or severely restricted their use in recent years [5].

However, these chemicals can persist in the environment for decades [4] and may still be found in some food and water sources.

Exposure to OCPs has been linked to a variety of health issues. OCPs have been linked to an increased risk of breast [6] and prostate [7] cancer, and lymphoma [8]. OCPs have been linked to reproductive problems in both men and women, including decreased fertility [9,10] and miscarriages, particularly with respect to hexachlorocyclohexane (HCH) [11], and prenatal exposure to α-HCH, β-HCH, o,p′-dichlorodiphenyldichloroethane (DDD), and p,p′-DDD has been associated with an increased risk for neural tube birth defects [12]. OCPs have been linked to neurological problems, which they are thought to contribute to via their accumulation in the body over time, potentially leading to Parkinson's disease (HCH) [13] and Alzheimer's disease (p,p′-dichlorodiphenyldichloroethylene (DDE)) [14]. OCPs have been linked to immune system disorders, with dichlorodiphenyltrichloroethane (DDT) causing the dysregulation of interleukin 6 (IL-6), leading to chronic inflammation [15], and having been related to oxidative stress and immune suppression [16], thus increasing susceptibility to infections and other diseases. OCPs have been classed as endocrine disruptors that can cause developmental problems in children (DDT) [17] and reproductive problems in adults [9]. In addition, low doses of OCPs have been strongly linked to various chronic diseases including diabetes [18] and cardiovascular diseases [19].

OCPs may have several mechanisms of action that contribute to their reported diverse effects. OCPs function by binding to specific target sites in the body, including binding to and blocking the GABAA receptors of sodium channels in the brain [20]. By binding to these sites, they disrupt the normal functioning of the nervous system, leading to the overstimulation or inhibition of nerve impulses. OCPs can also interfere with the normal function of the endocrine system by disrupting the production, release, transport, and metabolism of hormones at several sites of action such as the hypothalamus (increasing gonadotropin-releasing hormone mRNA levels), pituitary glands (affecting luteinizing hormone release), and ovaries (decreased sex steroid hormone production) [20].

Vitamin D deficiency is associated with various health conditions such as osteoporosis, cancer, cardiovascular disease, autoimmune diseases, and increased mortality [21,22]. In addition, vitamin D deficiency is associated with an increased incidence of type 2 diabetes [23–25] and microvascular and macrovascular diabetes-related complications [26]. Vitamin D_3 (cholecalciferol, $25(OH)D_3$) is endogenously produced via the UV-B irradiation of 7-dehydrocholesterol and is hydroxylated to $25(OH)D_3$ by multiple 25-hydroxylases in the liver [27]. $25(OH)D_3$ is transported to the kidneys and converted to active 1,25-dihydrovitamin D_3 ($1,25(OH)_2D_3$) by 1 alpha hydroxylase present in the renal tubular cells [28]. It has recently been reported that extrarenal tissues may also convert $25(OH)D_3$ to $1,25(OH)_2D_3$, although activation in renal and non-kidney tissues is regulated differently via the macrophage production of $1,25(OH)_2D_3$ through the type 2 interferon (IFN) response [29].

There is some evidence suggesting that exposure to OCPs may be associated with lower levels of vitamin D in the body [19,30]. Population studies have suggested that OCPs, including p,p′-DDT, p,p′-DDE, and β-HCH, can cause vitamin D deficiency [31]. In one such study, Inuit people who were heavily exposed to these pesticides through their traditional diet of fish and marine mammals, causing accumulation of these pesticides, showed higher levels of OCPs in their blood and had lower levels of vitamin D [32]; however, chronobiological factors within this population are likely to play a role.

OCPs may contribute to vitamin D deficiency through their effects on vitamin D metabolism in the liver, the disruption of the conversion of $25(OH)D_3$ to $1,25(OH)_2D_3$, interference with the function of vitamin D receptors in the body, or the impairment of calcium homeostasis [33]. It is important to note that the evidence linking OCPs and vitamin D levels is still limited, and more research is needed to fully understand the relationship between these two factors. However, the evidence still highlights the potential health risks associated with exposure to these persistent environmental pollutants. Therefore, this study was undertaken to examine the association between OCPs (pentachlorobenzene (PeCB), hexachlorobenzene (HCB), HCH, lindane, trans-chlordane, cis-chlordane, DDE, DDD, DDT, and dodecachlorooctahydro-1,3,4-metheno-1H-cyclobuta[cd]pentalene (Mirex)) and 25(OH)D_3 and its active metabolite 1,25$(OH)_2D_3$ in a group of non-obese healthy women prior to undergoing in vitro fertilization (IVF). In addition, we sought to determine if such an association may be precipitated by the impairment of calcium homeostasis, specifically with respect to its association with calcium/calmodulin-dependent (CaM) kinases (CaMK), CaMK kinase alpha (CaMKKα), and CaM phosphodiesterase (PDE) proteins.

2. Materials and Methods

2.1. Patient Recruitment

Subjects were recruited from the Hull IVF Unit, UK, following ethical approval from The Yorkshire and The Humber NRES ethical committee, UK (approval number 02/03/043). Venesection was performed on 58 fasting, non-obese Caucasian women; samples were taken at 09:00 on day 21 of the menstrual cycle prior to IVF downregulation. Venous blood samples were centrifuged at 3500× *g* for 15 min at 4 °C, and serum was stored at −80 °C within 1 h of collection for further analysis. No participant was on any prescribed or over-the-counter medication. All participants provided written informed consent.

2.2. OCP Measurement

Samples were screened for 10 indicator OCPs (PeCB, HCB, HCH, lindane, trans-chlordane, cis-chlordane, DDE, DDD, DDT, and Mirex). A sum-of-OCPs (ΣOCPs) variable was calculated by adding the molar concentrations of each OCP analyzed. After vortexing the samples for 1 min, 6 mL of acetonitrile, 3 mL of milliQ water, 5 g of anhydrous $MgSO_4$, and 1 g of NaCl were added along with a ceramic homogenizer. Samples were manually shaken for 1 min before being centrifuged at 4700 RPM for 8 min at 10 °C. The supernatant layer (approximately 6 mL of acetonitrile) was collected and transferred to 12 mL glass vials. An aliquot of 200 µL was collected and cleaned for non-targeted analysis (not presented here). The remainder of the extract was transferred to a 15 mL glass tube, evaporated under a gentle stream of nitrogen to near-dryness, and reconstituted in approximately 1 mL of hexane. Approximately 1 mL of >98% concentrated sulfuric acid was added, and the sample was vortexed for at least 30 s. The aqueous and hexane layers were left to separate overnight at <4 °C. The supernatant (hexane) layer was transferred directly onto a silica solid-phase extraction (SPE) cartridge (Supelco LC-Si, 3 mL/500 mg) (preconditioned with 6 mL dichloromethane (DCM) followed by 6 mL of hexane). Target compounds were eluted into a 15 mL glass collection tube using 6 mL of hexane followed by 8 mL of DCM. Samples were evaporated to near-dryness and reconstituted in 50 µL of methanol containing 2.5 ng $^{13}C_{12}$-PCB-141 as a recovery standard. Samples were transferred to autosampler vial inserts prior to analysis.

2.3. Instrumental Analysis

OCPs were identified in serum samples via high-resolution gas chromatography coupled with high-resolution mass spectrometry (HRGC/HRMS). A Thermofisher TRACE 1300 gas chromatograph was coupled to a Thermofisher DFS mass spectrometer. The injector was operated in splitless mode with separation achieved on an Agilent DB-5ms column (30 m length × 0.25 mm in diameter × 0.25 µm film thickness). Experiments were conducted in multiple ion detection (MID) mode at a resolution level of 10,000 (10% valley definition). The inlet, transfer line, and source were maintained at 250 °C, 280 °C, and 280 °C, respectively. The flow rate was maintained at 1.0 mL/min. In the GC oven, an initial temperature of 80 °C was held for two minutes before being ramped up to 180 °C at 20 °C/min and held for 0.5 min. The temperature was then increased to 300 °C at 10 °C/min and held for 5 min.

2.4. Quality Assurance/Quality Control

The limit of reporting (LOR) was determined according to the quality of the peak shape and using a minimum signal-to-noise ratio of 10:1. Contamination was assessed via inclusion of procedural blanks (milli-Q water, n = 6) extracted in parallel with unknown samples. If a target compound was detected in a blank at below 5% of a sample, then no correction occurred. When blank concentrations were 5–25% of a respective sample, the sample was blank corrected. Linearity was established with a seven-point calibration across the range of 0.1 to 25 ng/mL for all target analytes. R2 values > 0.99 were achieved for all compounds. Analytes were identified based on retention time (<2% RSD from standard) and confirmed with reference to a second ion (HRGC/HRMS). Quantification was performed using isotope dilution mass spectrometry. All OCPs were quantified against 13C12-HCB. All data were validated by spiking 5 aliquots of bovine serum with 300 pg of target analytes. The concentrations of the spiked samples were measured, and the recoveries (measured concentration/spiked concentration × 100) were calculated. Average recoveries were 80–120% with a relative standard deviation of <15%, thus conforming to the OECD guidelines of good laboratory practice (Supplementary Table S1).

2.5. Vitamin D_3 and Biochemical Parameters

Biochemical and hormonal parameters were measured as per a previously detailed method [34]. Briefly, serum 25(OH)D_3 and 1,25$(OH)_2D_3$ levels were quantified using isotope dilution liquid chromatography tandem mass spectrometry (LC-MS/MS). Vitamin D metabolites and labelled internal standards were extracted from 250 µL of serum using supportive liquid–liquid extraction and Diels–Alder derivatization prior to LC-MS/MS analysis. Chromatographic separations were achieved using Hypersil Gold C18 column (150 × 2.1 mm; 1.9 µm) at a flow rate of 0.2 mL/min and operated in Electrospray Ionization (ESI) positive mode and analyzed using the multiple-reaction-monitoring (MRM) method. The limit of quantification (LOQ) for 1,25$(OH)_2D_3$ was 10 pg/mL and 25(OH)D_3 1.2 nmol/L (0.5 ng/mL) [35]. Vitamin D deficiency was defined as a cutoff of 20 ng/mL (50 nmol/L) as per the clinical practice guidelines of the Endocrine Society Task Force on Vitamin D [36].

2.6. CaM Proteomic Measurement

Circulating levels of CaM-associated proteins (PDE1A, CaMK2A, CaMK2B, CaMK1D, CaMK2D, CaMK1, and CaMKKα) were determined via Slow Off-rate Modified Aptamer (SOMA)-scan plasma protein measurement (Somalogic, Boulder, CO, USA), the details of which have been previously reported [37]. Briefly, EDTA plasma samples were diluted, and the following assay steps were performed: (a) Binding: analytes and primer bead–SOMAmers (fully synthetic fluorophore-labelled SOMAmer coupled to a biotin moiety through a photocleavable linker) are equilibrated. (b) Catch I: all analyte–SOMAmer complexes are immobilized on a streptavidin-substituted support. Washing steps: proteins not stably bound to primer bead–SOMAmers are removed, and bound protein is biotinylated. (c) Cleave: long-wave ultraviolet light is applied to release analyte SOMAmer complexes into the solution. (d) Catch II: analyte–SOMAmer complexes are selectively immobilized on streptavidin support via the introduced analyte-borne biotinylation. Further washing is performed to select against non-specific analyte–SOMAmer complexes. (e) Elution: denaturation is induced to disrupt analyte–SOMAmer complexes. Released SOMAmers serve as surrogates for quantification of analyte concentrations. (f) Quantification: SOMAmer complementary oligonucleotides are hybridized to custom arrays. Normalization of raw intensities, hybridization, median signal, and calibration signal were performed based on the standard samples included on each plate, as previously described [38].

2.7. Statistics

No previous studies were available to enable the performance of a power analysis; therefore, this pilot study was designed according to the research conducted by Birkett and Day [39]. Descriptive data are presented as mean ± standard deviation (SD) for continuous data. Potential correlations between OCPs, 25(OH)D_3), and 1,25$(OH)_2D_3$ were examined using exploratory Spearman's rank order correlations. A *p*-value of <0.05 was considered to indicate statistical significance. Statistical analysis was carried out using Jamovi (version 2.0.0).

3. Results

3.1. Whole-Cohort Analysis

Demographic and biochemical data for this cohort are shown in Table 1. The participants had a mean age of 31.9 ± 4.6 years and a mean body mass index (BMI) of 25.7 ± 3.7 kg/m^2. Thyroid function and C-reactive protein (CRP) levels, constituting measures of underlying inflammation, were normal. The mean levels of 25(OH)D_3 and 1,25$(OH)_2D_3$ were 22.9 ± 11.2 ng/mL and 0.05 ± 0.02 ng/mL, respectively. Of the 58 women recruited, 28 had a 25(OH)D_3 level less than 20 ng/mL (50 nmol/L). The obtained levels of OCPs, 25(OH)D_3, and 1,25$(OH)_2D_3$ are shown in Table 1.

Table 1. Mean and standard deviation of OCPs in the population studied.

	Female Subjects (n = 58)	
	Mean	**SD**
Age (years)	31.9	4.6
BMI (kg/m^2)	25.7	3.7
CRP (mg/L)	2.6	2.5
TSH (mU/L)	2.4	2.2
Free-T3 (pmol/L)	4.8	0.7
Free-T4 (pmol/L)	11.3	1.8
25(OH)D_3 (ng/mL)	23	11.2
1,25$(OH)_2D_3$ (ng/mL)	0.05	0.02
Calcium (mmol/L)	2.3	0.07
Urea (nmol/L)	4.66	5.21
Creatinine (nmol/L)	66	8.86
eGFR (mL/min/1.73 m^2)	92.3	14.8
PDE1A (RFU)	789	1060
CAMK2A (RFU)	373	453
CaMK2B (RFU)	859	1423
CaMK1D (RFU)	1695	580
CaMK2D (RFU)	2379	3462
CaMK1 (RFU)	4681	1371
CaMKK α (RFU)	343	463
PeCB (ng/g Lipid)	9.12	9.1
α-HCH (ng/g Lipid)	<LOR	
β-HCH (ng/g Lipid)	4.63	4.92
HCB (ng/g Lipid)	9.81	6.5
Lindane (ng/g Lipid)	1.35	0.28
Trans-Chlordane (ng/g Lipid)	1.4	0.97
cis-Chlordane (ng/g Lipid)	1.59	0.58
p,p′-DDE (ng/g Lipid)	40.7	60.4
o,p′-DDE (ng/g Lipid)	2.23	0.32
o,p′-DDD (ng/g Lipid)	<LOR	
p,p′-DDD+o,p′-DDT (ng/g Lipid)	<LOR	
p,p′-DDT (ng/g Lipid)	4.99	2.3
Mirex (ng/g Lipid)	3.27	0.69
ΣOCPs (ng/g Lipid)	63.3	66.8

BMI: body mass index; CRP: C-reactive protein; TSH: thyroid-stimulating hormone; Free-T3: free-triiodothyronine; Free-T4: free-thyroxine; eGFR: estimated glomerular filtration rate; PDE: calcium/calmodulin-dependent phosphodiesterase; CaMK: calcium/calmodulin-dependent kinase; CaMKK α: calcium/calmodulin-dependent kinase kinase alpha; PeCB pentachlorobenzene; α: alpha; β: beta; HCH: hexachlorocyclohexane; HCB: hexachlorobenzene; p: para; o: ortho; DDE: dichlorodiphenyldichloroethylene; DDD: dichlorodiphenyldichloroethane; DDT: dichlorodiphenyltrichloroethane; Mirex: dodecachlorooctahydro-1,3,4-metheno-1H-cyclobuta[cd]pentalene; SD: standard deviation; LOR: limit of reporting; ΣOCPs: total organochlorine pesticides.

3.2. OCP Levels

The detection frequency of individual OCPs ranged from 100–12%. p,p′-DDE was detected in all serum samples, making it the most abundant OCP, followed by HCB (95%), p,p′-DDT (74%), β-HCH (67%), cis-chlordane (48%), Mirex (41%), PeCB (38%), trans-chlordane (24%), o,p′-DDE (22%), and lindane (12%). The OCPs α-HCH, o,p′-DDD, and p,p′-DDD+o,p′-DDT were not detected.

3.3. Whole-Group Correlations

In the study cohort, no correlations were found between ΣOCPs or individual OCPs and 25(OH)D_3. The OCP p,p′-DDE and the ΣOCPs positively correlated with 1,25$(OH)_2D_3$, (ρ = 0.41; p = 0.01) and (ρ = 0.34; p = 0.04), respectively (Figure 1A,B), but no other correlations were found between individual OCPs and 1,25$(OH)_2D_3$ (Table 2). The ΣOCPs variable negatively correlated with eGFR (ρ = −0.28; p = 0.04) (Figure 1C). There were no correlations between p,p′-DDE or ΣOCPs with urea and creatinine.

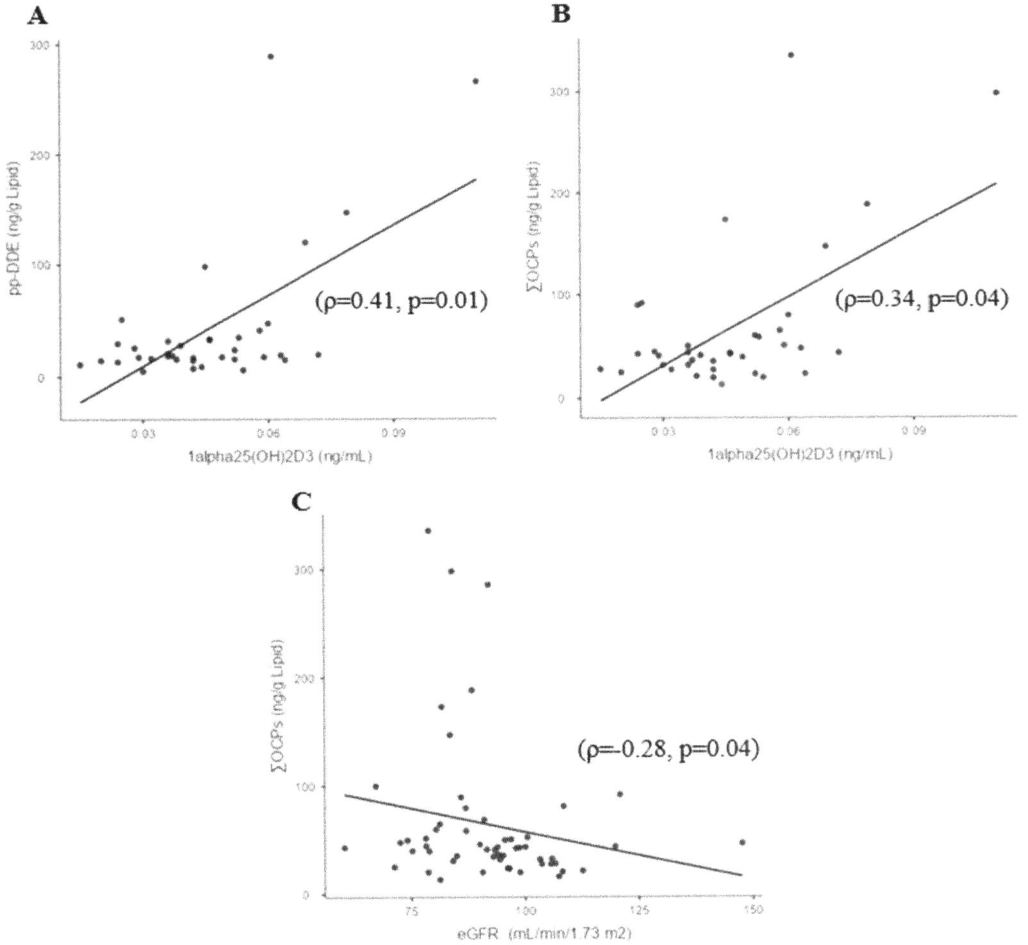

Figure 1. Significant correlations of OCPs with 1,25$(OH)_2D_3$ and measures of renal function in whole study cohort. (**A**) 1,25$(OH)_2D_3$ and p,p′-DDE; (**B**) 1,25$(OH)_2D_3$ and ΣOCPs; (**C**) eGFR and ΣOCPs.

Table 2. Spearman's rho (ρ) correlations of OCPs with $25(OH)D_3$ and $1,25(OH)_2D_3$. Data in bold indicate statistically significant correlations.

	PeCB	β-HCH	HCB	Lindane	trans-Chlordane	cis-Chlordane	o,p′-DDE	p,p′-DDE	p,p′-DDT	Mirex	ΣOCPs
$25(OH)D_3$	0.289 (0.192)	0.174 (0.339)	0.101 (0.502)	0.393 (0.40)	−0.14 (0.648)	0.325 (0.091)	0.374 (0.258)	0.272 (0.061)	0.096 (0.595)	−0.204 (0.361)	0.284 (0.051)
$1,25(OH)_2D_3$	−0.108 (0.66)	0.315 (0.14)	0.333 (0.05)	−0.2 (0.92)	−0.205 (0.74)	0.278 0.263	0.06 (0.89)	**0.408 (0.01)**	0.364 (0.09)	−0.03 (0.92)	**0.34 (0.04)**

3.4. Sufficient and Deficient 25(OH)D3 Subgroup Correlations

The study cohort was divided into two subgroups based on the participants' vitamin D levels: sufficient $25(OH)D_3$ (>20 ng/mL; >50 nmol/L) women and deficient $25(OH)D_3$ (<20 ng/mL; <50 nmol/L) women. In the sufficient $25(OH)D_3$ women, p,p′-DDT strongly positively correlated with $1,25(OH)_2D_3$ (ρ = 0.70; p = 0.01) (Figure 2A). There were no correlations found between OCPs and $25(OH)D_3$ in the sufficient $25(OH)D_3$ women subgroup. In deficient $25(OH)D_3$ women, HCB and p,p′-DDE positively correlated with $1,25(OH)_2D_3$ ((ρ = 0.55; p = 0.03) and (ρ = 0.52; p = 0.045), respectively (Figure 2B,C)), while β-HCH strongly positively correlated with $25(OH)D_3$ (ρ = 0.85; p = 0.01) (Figure 2D). Calcium levels did not differ between vitamin D sufficiency (2.3 ± 0.07 mmol/L) and vitamin D deficiency groups (2.3 ± 0.06 mmol/L).

We have found that $1,25(OH)_2D_3$ levels may be modulated through CaMK1 by polybrominated diethyl ether 153 (unpublished results); therefore, the CaM-associated proteins (PDE1A, CaMK2A, CaMK2B, CaMK1D, CaMK2D, CaMK1, and CaMKKα) (Table 1) that effect the rapid action of $1,25(OH)_2D_3$ at the cell membrane leading to increasing intracellular calcium levels were investigated, but no associations with the individual or total OCPs, or in vitamin-D-sufficient or deficient states, were found.

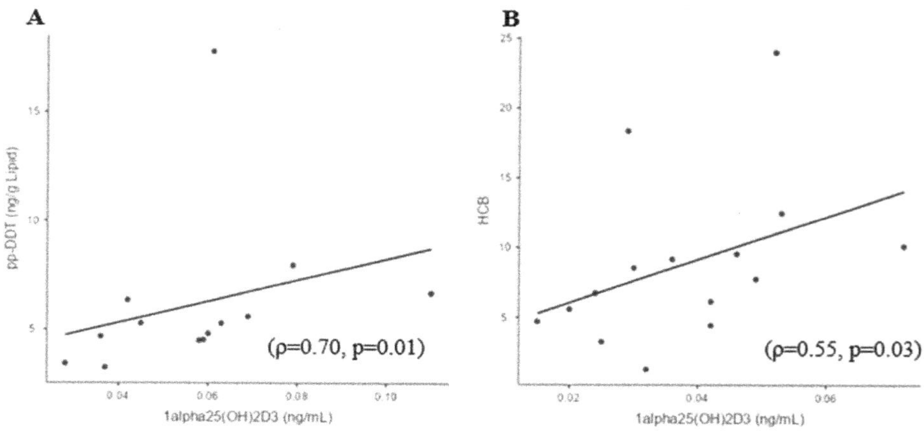

Figure 2. *Cont.*

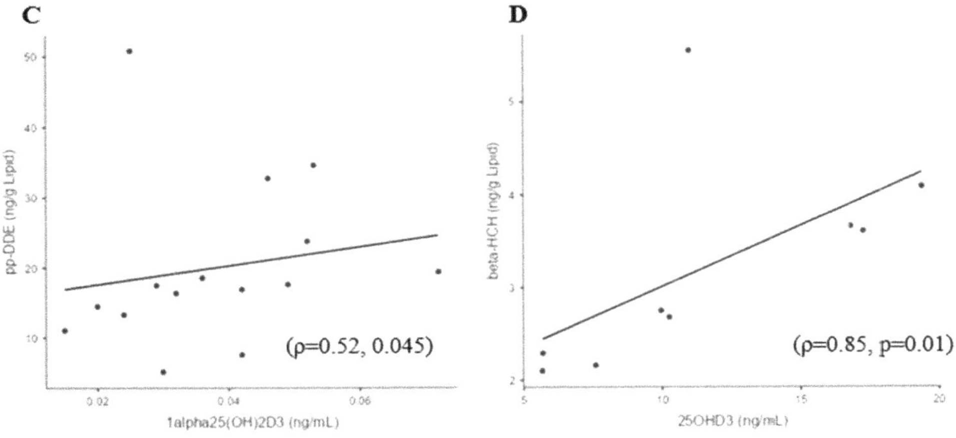

Figure 2. Significant correlations of OCPs with $25(OH)D_3$ and $1,25(OH)_2D_3$ in cohort subgroups. The study cohort was divided into two subgroups based on the participants' vitamin D levels: sufficient $25(OH)D_3$ (>20 ng/mL; >50 nmol/L) women and deficient $25(OH)D_3$ (<20 ng/mL; >50 nmol/L) women. Sufficient vitamin D subgroup—(**A**) $1,25(OH)_2D_3$ and p,p′-DDT. Deficient vitamin D subgroup—(**B**) $1,25(OH)_2D_3$ and HCB; (**C**) $1,25(OH)_2D_3$ and p,p′-DDE; (**D**) $25(OH)D_3$ and β-HCH.

4. Discussion

These data show that the 10 indicator OCPs and ΣOCPs analyzed do not appear to be associated with $25(OH)D_3$ but that the OCP p,p′-DDE and ΣOCPs correlated positively with active $1,25(OH)_2D_3$. Overall, the ΣOCPs were negatively associated with a reduced eGFR level, as noted by others [40], indicating that their action may be exerted on the kidneys, as suggested by the reported association of DDT burden with chronic kidney disease [41], but this was an observational study with no mechanistic insights. An inverse relationship of OCPs, including p,p′-DDT, p,p′-DDE, and β-HCH, with vitamin D has been reported [31], but this was not seen in the data presented herein; however, this difference is potentially due to the differing study populations, as those associations were seen amongst subjects of an older age, who were Caucasian, or who suffered from chronic diseases, whereas the population analyzed in this study was young and healthy.

When the study cohort was divided into two subgroups, namely, sufficient and deficient groups based on their vitamin D levels, in the vitamin-D-sufficient $25(OH)D_3$ women, p,p′-DDT strongly positively correlated with $1,25(OH)_2D_3$. There were no correlations found between OCPs and $25(OH)D_3$ in the vitamin-D-sufficient $25(OH)D_3$ subgroup of women. In the vitamin-D-deficient $25(OH)D_3$ women, both p,p′-DDE and HCB positively correlated with $1,25(OH)_2D_3$. In the vitamin-D-deficient group, β-HCH strongly positively correlated with $25(OH)D_3$, suggesting that a reduction in vitamin D may signal a potential effect of both HCB and HCH, but it is unclear what those specific effects may be. The mechanism(s) behind the positive associations between OCPs and $1,25(OH)_2D_3$ is unclear, but it has been reported that in cells exposed to p,p′-DDT and p,p′-DDE, this led to an increase in the vitamin D binding protein [42] that binds to $1,25(OH)_2D_3$ for its activity, perhaps leading to an increase in $1,25(OH)_2D_3$ levels.

$1,25(OH)_2D_3$ increases calcium absorption via vitamin D receptor (VDR) binding and exerts rapid action at the cell membrane, increasing intracellular calcium levels by activating protein kinase cascades that stimulate VDR-mediated transcription-enhancing steroid receptor coactivator (SRC) activity, with the overall effect of enhancing $1,25(OH)_2D_3$-dependent interactions between VDR and SRC coactivators [43]. To investigate a possible mechanism by which OCPs may affect $1,25(OH)_2D_3$ action, the determination of the plasma levels of CaM proteins (CaMK, CaMKK, and PDE), which may be indicative of

1,25(OH)$_2$D$_3$ action, was undertaken. No correlations with total or individual OCPs were found, suggesting that the OCPs impacting 1,25(OH)$_2$D$_3$ are not acting at the membrane level through calmodulin-associated proteins. Therefore, it is clear that 25(OH)D$_3$ is converted to its biologically active form 1,25(OH)$_2$D$_3$ in the presence of OCPs without affecting serum calcium levels, as has been described in other animal species [44], and it is unclear how OCPs may then impair the vitamin D biological response. That the OCPs appeared to be associated with vitamin D and vitamin D sufficiency or deficiency suggests that OCPs' effects may be mediated through vitamin D. Of particular concern, OCPs may potentiate vitamin D deficiency, which is a condition that affects over 50% of the global population [45]. Moreover, the diseases that are associated with vitamin D deficiency, such as diabetes and cardiovascular disease [21,22], are the same as those attributed to OCP exposure [19,30], suggesting that one may represent an epiphenomenon of the other. Another potential explanation is that the association between OCPs and vitamin D may be caused by indirect mechanisms. In vitro, OCPs induce inflammation and oxidative stress, with β-HCH, DDE, and Dieldrin causing the upregulation of pro-inflammatory cytokines such as tumor necrosis factor (TNF)-α, IL-1β, IL-6, nuclear factor kappa B (NF-kB), and cyclooxygenase (COX)-2 expression in OCP-treated human ovary surface epithelial cells [46]. These findings are in line with observations that DDT causes the dysregulation of IL-6 leading to chronic inflammation [15] and that DDT is related to oxidative stress and immune suppression [15,16]. This is also reflected in in vivo findings, where, in a systematic review, exposure to OCPs was positively associated with CRP, IL-1β, IL-2, and IL-10, which is indicative of dysregulated inflammatory responses; however, the underlying mechanisms remain unclear [47]. Conversely, vitamin D prevents oxidative stress, while vitamin D deficiency induces the disruption of mitochondrial function [48] and is also associated with markers of inflammation [49]. Consequently, in vitamin D sufficiency, this may generally mitigate the inflammatory and oxidative stress effects of the OCPs; however, in vitamin D deficiency, these effects have already been uncovered and indeed may be exacerbated (Figure 3), as seen with results reporting that vitamin D deficiency promotes colonic inflammation [50] and, in an animal model of acetaminophen-induced liver injury, exacerbates induced oxidative stress [51].

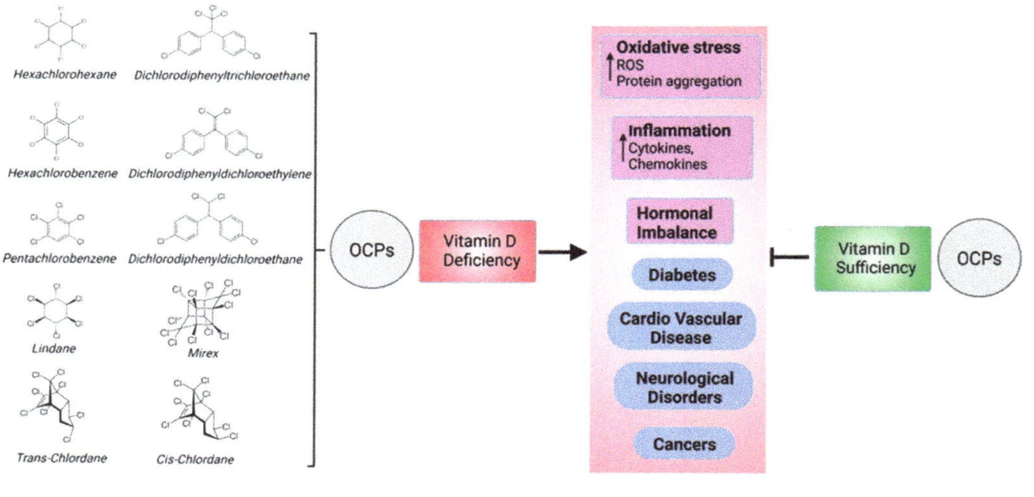

Figure 3. OCPs analyzed in this study. The association of OCPs with various diseases in vitamin D sufficiency and deficiency were analyzed. Vitamin D sufficiency may generally mitigate the inflammatory and oxidative stress effects and associated disorders/diseases arising due to exposure to OCPs, but in vitamin D deficiency, those effects are unmasked and aggravated.

In addition, extrarenal tissues may also convert 25(OH)D_3 to 1,25$(OH)_2D_3$; however, notably, activation in renal and non-kidney tissues is regulated differently, with macrophage production of 1,25$(OH)_2D_3$ through the type 2 IFN response [29]. In animal models, OCPs do not appear to affect macrophages; therefore, their influence through an extrarenal effect on 1,25$(OH)_2D_3$ appears less likely [52]. However, DDT caused an increase in IFNγ secretion [53]; increasing IFNγ may have the effect of increasing the conversion of 25(OH)D_3 to 1,25$(OH)_2D_3$ in the extrarenal tissues.

The strengths of this study include its state-of-the-art measurement of OCPs and vitamin D (25(OH)D_3 and 1,25$(OH)_2D_3$). Its limitations include the small number of subjects included, who were all Caucasian females, so these findings may not be generalizable to male subjects or those of different ethnicities. The low numbers compounded by the detection number may have resulted in a type-two statistical error (false negative); however, this study will potentiate the determinative power of a larger study focusing on vitamin D deficiency. OCPs have been associated with parathyroid hyperplasia leading to elevated parathyroid hormone (PTH) and calcium levels [54]; however, the calcium levels were in the normal reference range in these patients, thus obviating the need to measure PTH.

5. Conclusions

In conclusion, our findings show that, overall, OCPs and ΣOCPs were not associated with 25(OH)D_3, suggesting that they are unrelated to vitamin D deficiency, and that the OCP p,p′-DDE and the ΣOCPs correlated positively with active 1,25$(OH)_2D_3$ and negatively with eGFR, suggesting a renal effect. Vitamin D deficiency unmasked an association between β-HCH and 25(OH)D_3 and between HCB and p,p′-DDE and 1,25$(OH)_2D_3$, suggesting that OCPs' effects may be enhanced in patients with vitamin D deficiency.

Supplementary Materials: The following supporting information can be downloaded at: https://www.mdpi.com/article/10.3390/biomedicines11051451/s1. Supplementary Table S1. OCP validation data.

Author Contributions: Conceptualization, S.L.A. and T.S.; methodology, S.L.A. and T.S.; validation, S.L.A.; formal analysis, E.B. and A.E.B.; investigation, D.S.D.; resources, S.L.A.; data curation, S.L.A.; writing—original draft preparation, E.B., A.E.B. and S.L.A.; writing—review and editing, E.B., A.E.B., S.L.A., D.S.D., M.N. and T.S.; visualization, E.B., A.E.B., M.N. and S.L.A. All authors have read and agreed to the published version of the manuscript.

Funding: This research received no external funding.

Institutional Review Board Statement: The study was conducted in accordance with the Declaration of Helsinki, and approved by the Institutional Review Board (or Ethics Committee) of The Yorkshire and The Humber NRES ethical committee, UK (approval number 02/03/043).

Informed Consent Statement: Informed consent was obtained from all subjects involved in the study.

Data Availability Statement: The data presented in this study are available on reasonable request from the corresponding author.

Conflicts of Interest: The authors declare no conflict of interest.

References

1. Jayaraj, R.; Megha, P.; Sreedev, P. Organochlorine pesticides, their toxic effects on living organisms and their fate in the environment. *Interdiscip. Toxicol.* **2016**, *9*, 90–100. [CrossRef] [PubMed]
2. Deshmukh, H.; Aylward, L.L.; Rose, M.; Fernandes, A.; Sedman, P.; Thatcher, N.J.; Atkin, S.L.; Sathyapalan, T. Association of endocrine active environmental compounds with body mass index and weight loss following bariatric surgery. *Clin. Endocrinol.* **2020**, *93*, 280–287. [CrossRef] [PubMed]
3. Buah-Kwofie, A.; Humphries, M.S.; Combrink, X.; Myburgh, J.G. Accumulation of organochlorine pesticides in fat tissue of wild Nile crocodiles (*Crocodylus niloticus*) from iSimangaliso Wetland Park, South Africa. *Chemosphere* **2018**, *195*, 463–471. [CrossRef] [PubMed]
4. Ma, Y.; Yun, X.; Ruan, Z.; Lu, C.; Shi, Y.; Qin, Q.; Men, Z.; Zou, D.; Du, X.; Xing, B.; et al. Review of hexachlorocyclohexane (HCH) and dichlorodiphenyltrichloroethane (DDT) contamination in Chinese soils. *Sci. Total Environ.* **2020**, *749*, 141212. [CrossRef]

5. Keswani, C.; Dilnashin, H.; Birla, H.; Roy, P.; Tyagi, R.K.; Singh, D.; Rajput, V.D.; Minkina, T.; Singh, S.P. Global footprints of organochlorine pesticides: A pan-global survey. *Environ. Geochem. Health* **2022**, *44*, 149–177. [CrossRef] [PubMed]
6. Miao, Y.; Rong, M.; Li, M.; He, H.; Zhang, L.; Zhang, S.; Liu, C.; Zhu, Y.; Deng, Y.L.; Chen, P.P.; et al. Serum concentrations of organochlorine pesticides, biomarkers of oxidative stress, and risk of breast cancer. *Environ. Pollut.* **2021**, *286*, 117386. [CrossRef]
7. Antignac, J.P.; Figiel, S.; Pinault, M.; Blanchet, P.; Bruyère, F.; Mathieu, R.; Lebdai, S.; Fournier, G.; Rigaud, J.; Mahéo, K.; et al. Persistent organochlorine pesticides in periprostatic adipose tissue from men with prostate cancer: Ethno-geographic variations association with disease aggressiveness. *Environ. Res.* **2023**, *216*, 114809. [CrossRef] [PubMed]
8. Luo, D.; Zhou, T.; Tao, Y.; Feng, Y.; Shen, X.; Mei, S. Exposure to organochlorine pesticides and non-Hodgkin lymphoma: A meta-analysis of observational studies. *Sci. Rep.* **2016**, *6*, 25768. [CrossRef]
9. Rattan, S.; Zhou, C.; Chiang, C.; Mahalingam, S.; Brehm, E.; Flaws, J.A. Exposure to endocrine disruptors during adulthood: Consequences for female fertility. *J. Endocrinol.* **2017**, *233*, R109–R129. [CrossRef]
10. Giulioni, C.; Maurizi, V.; Castellani, D.; Scarcella, S.; Skrami, E.; Balercia, G.; Galosi, A.B. The environmental and occupational influence of pesticides on male fertility: A systematic review of human studies. *Andrology* **2022**, *10*, 1250–1271. [CrossRef]
11. Pathak, R.; Mustafa, M.; Ahmed, R.S.; Tripathi, A.K.; Guleria, K.; Banerjee, B.D. Association between recurrent miscarriages and organochlorine pesticide levels. *Clin. Biochem.* **2010**, *43*, 131–135. [CrossRef] [PubMed]
12. Yin, S.; Sun, Y.; Yu, J.; Su, Z.; Tong, M.; Zhang, Y.; Liu, J.; Wang, L.; Li, Z.; Ren, A.; et al. Prenatal exposure to organochlorine pesticides is associated with increased risk for neural tube defects. *Sci. Total Environ.* **2021**, *770*, 145284. [CrossRef]
13. Xu, S.; Yang, X.; Qian, Y.; Luo, Q.; Song, Y.; Xiao, Q. Analysis of serum levels of organochlorine pesticides and related factors in Parkinson's disease. *Neurotoxicology* **2022**, *88*, 216–223. [CrossRef] [PubMed]
14. Medehouenou, T.C.M.; Ayotte, P.; Carmichael, P.H.; Kröger, E.; Verreault, R.; Lindsay, J.; Dewailly, É.; Tyas, S.L.; Bureau, A. Laurin, D. Exposure to polychlorinated biphenyls and organochlorine pesticides and risk of dementia, Alzheimer's disease and cognitive decline in an older population: A prospective analysis from the Canadian Study of Health and Aging. *Environ. Health* **2019**, *18*, 57. [CrossRef] [PubMed]
15. Martin, T.J.; Gabure, S.; Maise, J.; Snipes, S.; Peete, M.; Whalen, M.M. The organochlorine pesticides pentachlorophenol and dichlorodiphenyltrichloroethane increase secretion and production of interleukin 6 by human immune cells. *Environ. Toxicol. Pharmacol.* **2019**, *72*, 103263. [CrossRef]
16. Koner, B.C.; Banerjee, B.D.; Ray, A. Organochlorine pesticide-induced oxidative stress and immune suppression in rats. *Indian J. Exp. Biol.* **1998**, *36*, 395–398.
17. Jeddy, Z.; Kordas, K.; Allen, K.; Taylor, E.V.; Northstone, K.; Dana Flanders, W.; Namulanda, G.; Sjodin, A.; Hartman, T.J. Prenatal exposure to organochlorine pesticides and early childhood communication development in British girls. *Neurotoxicology* **2018**, *69*, 121–129. [CrossRef]
18. Evangelou, E.; Ntritsos, G.; Chondrogiorgi, M.; Kavvoura, F.K.; Hernández, A.F.; Ntzani, E.E.; Tzoulaki, I. Exposure to pesticides and diabetes: A systematic review and meta-analysis. *Environ. Int.* **2016**, *91*, 60–68. [CrossRef]
19. Zago, A.M.; Faria, N.M.X.; Fávero, J.L.; Meucci, R.D.; Woskie, S.; Fassa, A.G. Pesticide exposure and risk of cardiovascular disease: A systematic review. *Glob. Public Health* **2022**, *17*, 3944–3966. [CrossRef]
20. Costa, L.G. The neurotoxicity of organochlorine and pyrethroid pesticides. *Handb. Clin. Neurol.* **2015**, *131*, 135–148. [CrossRef]
21. Bjelakovic, G.; Gluud, L.L.; Nikolova, D.; Whitfield, K.; Wetterslev, J.; Simonetti, R.G.; Bjelakovic, M.; Gluud, C. Vitamin D supplementation for prevention of mortality in adults. *Cochrane Database Syst. Rev.* **2014**, Cd007470. [CrossRef] [PubMed]
22. Osorio Landa, H.K.; Perez Diaz, I.; Laguna Barcenas, S.D.C.; Lopez Navarro, J.M.; Abella Roa, M.F.; Corral Orozco, M.; Mancilla Ortega, J.P.; Martinez Duarte, D.A.; Morales Montalvo, S.I.; Muzquiz Aguirre, S.; et al. Association of serum vitamin D levels with chronic disease and mortality. *Nutr. Hosp.* **2020**, *37*, 335–342. [CrossRef] [PubMed]
23. Nakashima, A.; Yokoyama, K.; Yokoo, T.; Urashima, M. Role of vitamin D in diabetes mellitus and chronic kidney disease. *World J. Diabetes* **2016**, *7*, 89–100. [CrossRef] [PubMed]
24. Husemoen, L.L.; Thuesen, B.H.; Fenger, M.; Jorgensen, T.; Glumer, C.; Svensson, J.; Ovesen, L.; Witte, D.R.; Linneberg, A. Serum 25(OH)D and type 2 diabetes association in a general population: A prospective study. *Diabetes Care* **2012**, *35*, 1695–170[illegible] [CrossRef] [PubMed]
25. Mathieu, C. Vitamin D and diabetes: Where do we stand? *Diabetes Res. Clin. Pract.* **2015**, *108*, 201–209. [CrossRef]
26. Butler, A.E.; Dargham, S.R.; Latif, A.; Mokhtar, H.R.; Robay, A.; Chidiac, O.M.; Jayyousi, A.; Al Suwaidi, J.; Crystal, R.G.; Abi Khalil, C.; et al. Association of vitamin D3 and its metabolites in subjects with and without Type 2 diabetes and their relationship to diabetes complications. *Ther. Adv. Chronic Dis.* **2020**, *11*, 2040622320924159. [CrossRef]
27. Bikle, D.D. Vitamin D metabolism, mechanism of action, and clinical applications. *Chem. Biol.* **2014**, *21*, 319–329. [CrossRef]
28. Christakos, S.; Dhawan, P.; Verstuyf, A.; Verlinden, L.; Carmeliet, G. Vitamin D: Metabolism, Molecular Mechanism of Action, and Pleiotropic Effects. *Physiol. Rev.* **2016**, *96*, 365–408. [CrossRef]
29. Adams, J.S.; Rafison, B.; Witzel, S.; Reyes, R.E.; Shieh, A.; Chun, R.; Zavala, K.; Hewison, M.; Liu, P.T. Regulation of the extrarenal CYP27B1-hydroxylase. *J. Steroid Biochem. Mol. Biol.* **2014**, *144 Pt A*, 22–27. [CrossRef]
30. Han, X.; Zhang, F.; Meng, L.; Xu, Y.; Li, Y.; Li, A.; Turyk, M.E.; Yang, R.; Wang, P.; Zhang, J.; et al. Exposure to organochlorine pesticides and the risk of type 2 diabetes in the population of East China. *Ecotoxicol. Environ. Saf.* **2020**, *190*, 110125. [CrossRef]
31. Yang, J.H.; Lee, Y.M.; Bae, S.G.; Jacobs, D.R., Jr.; Lee, D.H. Associations between organochlorine pesticides and vitamin D deficiency in the U.S. population. *PLoS ONE* **2012**, *7*, e30093. [CrossRef] [PubMed]

32. Schæbel, L.K.; Bonefeld-Jørgensen, E.C.; Vestergaard, H.; Andersen, S. The influence of persistent organic pollutants in the traditional Inuit diet on markers of inflammation. *PLoS ONE* **2017**, *12*, e0177781. [CrossRef] [PubMed]
33. Andrews, J.E.; Courtney, K.D.; Donaldson, W.E. Impairment of calcium homeostasis by hexachlorobenzene (HCB) exposure in Fischer 344 rats. *J. Toxicol. Environ. Health* **1988**, *23*, 311–320. [CrossRef] [PubMed]
34. Brennan, E.; Kumar, N.; Drage, D.S.; Cunningham, T.K.; Sathyapalan, T.; Mueller, J.F.; Atkin, S.L. A case-control study of polychlorinated biphenyl association with metabolic and hormonal outcomes in polycystic ovary syndrome. *J. Environ. Sci. Health C Toxicol. Carcinog.* **2022**, *40*, 86–105. [CrossRef] [PubMed]
35. Javed, Z.; Papageorgiou, M.; Deshmukh, H.; Kilpatrick, E.S.; Mann, V.; Corless, L.; Abouda, G.; Rigby, A.S.; Atkin, S.L.; Sathyapalan, T. A Randomized, Controlled Trial of Vitamin D Supplementation on Cardiovascular Risk Factors, Hormones, and Liver Markers in Women with Polycystic Ovary Syndrome. *Nutrients* **2019**, *11*, 188. [CrossRef]
36. Holick, M.F.; Binkley, N.C.; Bischoff-Ferrari, H.A.; Gordon, C.M.; Hanley, D.A.; Heaney, R.P.; Murad, M.H.; Weaver, C.M. Evaluation, treatment, and prevention of vitamin D deficiency: An Endocrine Society clinical practice guideline. *J. Clin. Endocrinol. Metab.* **2011**, *96*, 1911–1930. [CrossRef]
37. Moin, A.S.M.; Sathyapalan, T.; Butler, A.E.; Atkin, S.L. Vitamin D association with coagulation factors in polycystic ovary syndrome is dependent upon body mass index. *J. Transl. Med.* **2021**, *19*, 239. [CrossRef]
38. Moin, A.S.M.; Sathyapalan, T.; Butler, A.E.; Atkin, S.L. Coagulation factor dysregulation in polycystic ovary syndrome is an epiphenomenon of obesity. *Clin. Endocrinol.* **2023**, *98*, 796–802. [CrossRef]
39. Birkett, M.A.; Day, S.J. Internal pilot studies for estimating sample size. *Stat. Med.* **1994**, *13*, 2455–2463. [CrossRef]
40. Siddharth, M.; Datta, S.K.; Bansal, S.; Mustafa, M.; Banerjee, B.D.; Kalra, O.P.; Tripathi, A.K. Study on organochlorine pesticide levels in chronic kidney disease patients: Association with estimated glomerular filtration rate and oxidative stress. *J. Biochem. Mol. Toxicol.* **2012**, *26*, 241–247. [CrossRef]
41. Lv, J.; Guo, L.; Gu, Y.; Xu, Y.; Xue, Q.; Yang, X.; Wang, Q.N.; Meng, X.M.; Xu, D.X.; Pan, X.F.; et al. National temporal trend for organophosphate pesticide DDT exposure and associations with chronic kidney disease using age-adapted eGFR model. *Environ. Int.* **2022**, *169*, 107499. [CrossRef] [PubMed]
42. Pavlíková, N.; Daniel, P.; Šrámek, J.; Jelínek, M.; Šrámková, V.; Němcová, V.; Balušíková, K.; Halada, P.; Kovář, J. Upregulation of vitamin D-binding protein is associated with changes in insulin production in pancreatic beta-cells exposed to p,p′-DDT and p,p′-DDE. *Sci. Rep.* **2019**, *9*, 18026. [CrossRef]
43. Ellison, T.I.; Dowd, D.R.; MacDonald, P.N. Calmodulin-dependent kinase IV stimulates vitamin D receptor-mediated transcription. *Mol. Endocrinol.* **2005**, *19*, 2309–2319. [CrossRef] [PubMed]
44. Nowicki, H.G.; Myrtle, J.F.; Norman, A.W. Effects of organochlorine insecticides on metabolism of cholecalciferol (vitamin D 3) in Rachitic cockerel. *J. Agric. Food Chem.* **1972**, *20*, 380–384. [CrossRef] [PubMed]
45. Siddiqee, M.H.; Bhattacharjee, B.; Siddiqi, U.R.; MeshbahurRahman, M. High prevalence of vitamin D deficiency among the South Asian adults: A systematic review and meta-analysis. *BMC Public Health* **2021**, *21*, 1823. [CrossRef]
46. Shah, H.K.; Sharma, T.; Banerjee, B.D. Organochlorine pesticides induce inflammation, ROS production, and DNA damage in human epithelial ovary cells: An in vitro study. *Chemosphere* **2020**, *246*, 125691. [CrossRef]
47. Liu, Z.; Lu, Y.; Zhong, K.; Wang, C.; Xu, X. The associations between endocrine disrupting chemicals and markers of inflammation and immune responses: A systematic review and meta-analysis. *Ecotoxicol. Environ. Saf.* **2022**, *234*, 113382. [CrossRef]
48. Dzik, K.P.; Kaczor, J.J. Mechanisms of vitamin D on skeletal muscle function: Oxidative stress, energy metabolism and anabolic state. *Eur. J. Appl. Physiol.* **2019**, *119*, 825–839. [CrossRef]
49. Querfeld, U. Vitamin D and inflammation. *Pediatr. Nephrol.* **2013**, *28*, 605–610. [CrossRef]
50. Wei, X.; Li, X.; Du, J.; Ge, X.; Sun, Y.; Li, X.; Xun, Z.; Liu, W.; Wang, Z.Y.; Li, Y.C. Vitamin D Deficiency Exacerbates Colonic Inflammation Due to Activation of the Local Renin-Angiotensin System in the Colon. *Dig. Dis. Sci.* **2021**, *66*, 3813–3821. [CrossRef]
51. Wang, Y.Q.; Geng, X.P.; Wang, M.W.; Wang, H.Q.; Zhang, C.; He, X.; Liang, S.M.; Xu, D.X.; Chen, X. Vitamin D deficiency exacerbates hepatic oxidative stress and inflammation during acetaminophen-induced acute liver injury in mice. *Int. Immunopharmacol.* **2021**, *97*, 107716. [CrossRef] [PubMed]
52. Glinski, A.; Liebel, S.; Pelletier, È.; Voigt, C.L.; Randi, M.A.; Campos, S.X.; Oliveira Ribeiro, C.A.; Filipak Neto, F. Toxicological interactions of silver nanoparticles and organochlorine pesticides in mouse peritoneal macrophages. *Toxicol. Mech. Methods* **2016**, *26*, 251–259. [CrossRef] [PubMed]
53. Massawe, R.; Drabo, L.; Whalen, M. Effects of pentachlorophenol and dichlorodiphenyltrichloroethane on secretion of interferon gamma (IFNγ) and tumor necrosis factor alpha (TNFα) from human immune cells. *Toxicol. Mech. Methods* **2017**, *27*, 223–235. [CrossRef] [PubMed]
54. McMullin, J.L.; Codner, J.; Patel, S.G.; Sharma, J.; Hu, X.; Jones, D.P.; Weber, C.J.; Saunders, N.D. Environmental Chemicals and their Association with Hyperparathyroidism. *World J. Surg.* **2023**, *47*, 296–303. [CrossRef] [PubMed]

biomedicines MDPI

Article

Association of Blood Levels of Vitamin D and Its Binding Protein with Clinical Phenotypes of Multiple Sclerosis

Suhail Al-Shammri [1,2,*], Arpita Chattopadhyay [1], Magdy Girgis Hanah [2], Suhail Doi [3] and Abayomi Akanji [4]

1 Department of Medicine, Faculty of Medicine, Kuwait University, Kuwait City P.O. Box 24923, Kuwait
2 Neurology Unit, Department of Medicine, Mubarak al Kabeer Hospital, Ministry of Health, Jabriya 13001, Kuwait
3 Department of Population Medicine, Qatar University, Doha P.O. Box 2713, Qatar
4 Department of Medical Sciences, Frank Netter School of Medicine, Quinnipiac University, Hamden, CT 06518, USA
* Correspondence: suhail.alshammri@ku.edu.kw; Tel.: +965-6777-7719

Abstract: Background: Low vitamin D levels may synergize with changing levels of the vitamin D binding protein (DBP) to precipitate in the development and clinical progression of multiple sclerosis (MS). In this study, this hypothesis was explored in groups of Kuwaiti healthy controls and patients with different clinical phenotypes of MS. Methods: Fasting serum concentrations of 25-hydroxyvitamin D [25(OH)D] and DBP were measured in 146 healthy controls and 195 patients with MS. The latter were classified according to the duration, type, and onset of the disease and the mode of treatment. Factors such as relapse/remitting, and the use of nutritional supplements were also considered. Results: The DBP levels were significantly lower in the patients than in the controls. This was more evident in newly diagnosed drug-naïve patients than in those patients with more established MS. MS status and severity were negatively impacted by concurrently low levels of 25(OH)D and DBP. This was most clearly expressed in drug-naïve patients and in those with a disease in relapse. It was also established that the 25(OH)D level had a significant positive correlation with the duration of the disease. Conclusion: Lower levels of 25(OH)D and DBP appear to have a synergistic effect on MS status. This was most clearly demonstrated in patients who were newly diagnosed (drug-naïve) and in those patients who were in relapse.

Keywords: multiple sclerosis; vitamin D; 25-hydroxyvitamin D; vitamin D binding protein; Kuwait; hypovitaminosis D; relapsing remitting; drug-naïve patients

1. Introduction

Multiple sclerosis (MS) is, presumably, an autoimmune-mediated demyelinating and neurodegenerative disease of the central nervous system (CNS), which usually affects young adults and causes significant irreversible neurological disability [1]. It is widely believed that both genetic and environmental factors are involved in its etiopathogenesis [2]. Further research has implicated serum vitamin D status and its binding in the circulation to the vitamin D binding protein (DBP) and genetic polymorphisms related to the DBP in the pathogenesis of MS [3]. Increasing evidence suggests that vitamin D sufficiency and related sun exposure appear to lower the risk of developing multiple sclerosis (MS) [4], and in established cases reduce relapse [5], slow progression [6], and result in fewer new lesions on the brain magnetic resonance imaging (MRI) [7].

DBP is the major plasma carrier of vitamin D and its metabolites. It is responsible for transporting much of the liver metabolite, 25-hydroxyvitamin D [25(OH)D], in the circulation system, particularly to the kidneys for further conversion to the active form 1, 25 dihydroxyvitamin D [1,25(OH)$_2$D]. The latter ultimately accesses target tissues again via significant DBP transport [8]. The bioavailability of the free vitamin D metabolites is regulated by their affinity to the DBP [9]. Therefore, higher concentrations of the DBP may keep 25(OH)D and $1,25(OH)_2D$ within the circulation system, while lower levels of the free vitamin D might negatively affect the potential benefits of vitamin D physiologically and, specifically, in MS patients.

In addition to its role as a carrier protein, DBP is involved in the immune system, acting as both a chemotactic factor for neutrophils and a potent macrophage activating factor after its exposure to circulating B and T lymphocytes [8,9]. In addition, DBP functions as an actin scavenger by binding G-actin with high affinity and sequesters monomer actin released in the circulation as a result of injury and disease. Therefore, it may impact several conditions [8], possibly including MS.

The association of changing levels of DBP with the presence of MS is, however, controversial. Various studies have suggested that DBP levels are (i) low in MS patients' cerebrospinal fluid (CSF), especially in those patients with relapsing MS [10,11]; (ii) low in the serum of patients with MS compared to the serum of healthy controls [12], and the lowest in relapsing–remitting (RRMS) patients [13,14]; (iii) higher in the CSF of patients with secondary progressive MS (SPMS) [15]; (iv) not different in the serum levels in patients with stable or active disease compared to the serum levels of controls [16,17]; and (v) according to at least one study [18], higher in RRMS patients. The implications of low DBP's association with low circulating free vitamin D are unclear.

To address some of these controversies, we examined the associations of serum DBP with the clinical phenotypes of MS, independently and in conjunction with the serum 25(OH)D concentrations in a group of patients with MS matched with a healthy control population. The variables considered were age, gender, body mass index (BMI), duration of disease, and treatment modalities, including the use of nutritional supplements.

2. Materials and Methods

2.1. Study Design and Study Population

In this observational study, we recruited MS patients and healthy controls, approximately matching the two groups for age and sex [19], after receiving voluntary informed consent. The MS patients were diagnosed according to the McDonald criteria [20] and were followed up by our team of experienced neurologists at the national MS clinic at Mubarak Al-Kabeer Hospital in Kuwait. None of the patients had received corticosteroids during the month prior to the evaluation. The control subjects were selected from healthy individuals among the Kuwaiti population, who had no documented history of autoimmune, inflammatory, and neurologic diseases and no family history of MS. None of the controls were taking any medications. The female subjects (controls and patients) were not on estrogen supplementation.

Both groups of subjects were interviewed using a structured questionnaire to collect information about age, gender, body mass index (BMI), daily habits (including sun exposure, choice of routine outdoor dressing, routine diet, and outdoor physical activity), past medical history, and current and past medications (including dietary supplements). Moreover, the patients' medical records were reviewed for relevant past medical history, current medications, and evidence of other chronic medical disorders requiring long-term treatment.

Thereafter, the subjects (patients and controls) had careful physical examinations. The MS patients were sub-grouped into those with established long-standing (on drug treatment) disease and those with newly diagnosed (drug-naïve) disease. They were also assigned to the clinical phenotypes of RRMS, secondary progressive MS (SPMS), and primary progressive MS (PPMS) based on the McDonald criteria [20]. The disease stage, with or without associated disability, was determined using the expanded disability status scale (EDSS) [21].

2.2. Biochemical Analysis

Fasting blood specimens were collected from all subjects and immediately centrifuged, and the serum samples were stored at −40 °C until analyzed using validated kit ELISA tests for the levels of total 25(OH)D (Immune Diagnostic Systems, Bensheim, Germany) and DBP (Quantikine, R&D systems, Minneapolis, MN, USA). The kit manufacturer's protocols were strictly adhered to, and the assay performance characteristics (intra- and inter-assay CVs, recoveries, and analytical sensitivity and specificity) were within acceptable limits.

As per the recommendations of the clinical practice guidelines of the Endocrine Society Task Force [22], the cutoff points utilized for classification in this study were: serum 25(OH)D < 50 nmol/L—vitamin D deficiency; <25 nmol/L—severe vitamin D deficiency [23]. The other routine blood counts and clinical biochemical tests (liver and renal function) were carried out using routine Coulter Counter and Auto-analyzer (Beckman-Coulter DxC 800, Brea, CA, United States) techniques.

2.3. Statistical Analysis

Qualitative variables were reported as numbers and percentages, and quantitative variables as medians and interquartile ranges expressed as the 25th–75th percentile or mean and standard deviation (SD), as appropriate. The differences between the two independent groups were tested with the Mann–Whitney U test. The chi-square test and Student's *t*-test were used for a pairwise comparison. A multivariable logistic regression model was utilized to assess associations between serum levels of 25(OH)D and DBP and MS status. The Spearman rank correlation test was used to measure the degree of association between two variables. The ODD ratios in this study were adjusted for possible confounding factors (age, gender, BMI, duration of disease, and treatment modalities, including use of nutritional supplements, mode of dressing, and daily direct sunlight exposure). Statistical analyses were carried out using Statistical Package for Social Sciences, SPSS for Windows (IBM SPSS Statistics 22, IBM Corporation, Armonk, NY, USA, 2013). A p-value < 0.05 was considered significant.

2.4. Ethical Considerations

The study protocol and consent forms were approved by the Ethical Committee, Faculty of Medicine, Kuwait University (IORG0007925). The study was conducted in compliance with the international guidelines for human research protection, as described in the Declaration of Helsinki.

3. Results

The characteristics of the study population (patients and controls) are shown in Table 1.

Table 1. The characteristics of the study population (patients and controls) ([24]).

	Controls (n)	Patients with MS (n)			p *	p $	p #
		All	Established Patients	Newly Diagnosed (Drug-Naïve) Patients			
Number (n)	146	195	134	61			
Age (in years) mean (SD)	33.8 (9.5)	33.2 (10.4)	34.4 (11.0)	30.8 (8.3)	0.589 [b]	0.665 [b]	0.029 [b]
Gender					0.526 [a]	0.672 [a]	0.466 [a]
Male	52 (35.6)	76 (39.0)	51 (38.1)	25 (41.0)			
Female	94 (64.4)	119 (61.0)	83 (61.9)	36 (59.0)			
BMI (kg/m^2)	28.7 (10)	28.6 (12.5)	28.8 (11.5)	28.2 (10.5)	0.359 [c]	0.241 [c]	0.933 [c]
25–29.9 (overweight)	50 (36.8)	59 (31.9)	39 (30.5)	20 (35.1)			
>30 (Obese)	49 (36.0)	62 (33.5)	42 (32.8)	20 (35.1)			

Table 1. *Cont.*

	Controls (n)	Patients with MS (n)			p *	p $	p #
		All	Established Patients	Newly Diagnosed (Drug-Naïve) Patients			
Daily direct sunlight exposure (minutes)					<0.001 [a]	<0.001 [a]	0.008 [a]
<10	36 (26.1)	62 (31.8)	37 (27.6)	25 (41.0)			
10–15	32 (23.2)	93 (47.7)	72 (53.7)	21 (34.4)			
15–30	35 (25.4)	25 (12.8)	17 (12.7)	8 (13.1)			
>30	35 (23.4)	15 (7.7)	8 (6.0)	7 (11.5)			
Mode of dressing					0.654 [a]	0.980 [a]	0.026 [a]
Traditional/Western (Exposed Face/Arms	112 (83.0)	158 (81.0)	116 (86.6)	42 (68.9)			
Fully covered eyes/hands Exposed	23 (17.0)	37 (19.0)	18 (13.4)	19 (31.1)			
MS Subtype							
Relapsing–remitting (n, %)	-	166 (85.2)	105 (78.4)	61 (100)			
Secondary progressive (n, %)	-	26 (13.3)	26 (19.4)	-			
Primary progressive (n, %)	-	3 (1.5)	3 (2.2)	-			
Age at recruitment year (median *)	-	32 (27–38)	33 (27–41)	31 (26–36)			
Age at diagnosis, year (median *)	-	28 (21–34)	27 (20–33)	31 (26–36)			
Age at onset, year, (median *)	-	27 (20–32)	26 (19–32)	29 (22–33)			
Duration of disease from onset of symptoms, year (median *)	-	3 (1–8)	5 (2–10)	2 (1–7)			
EDSS score (median *)	-	1.5 (0–2.5)	1.5 (0–3)	1 (0–2)			
Annualised relapse rate (median *)		3 (1–5)	4 (2–7)	0.1 (0.1–0.2)			

n: number of subjects. Numeric data are expressed as the mean (SD), medians * (25th–75th percentile), and percentage (%), as appropriate; p-values are generated via [a] chi-square test, [b] Student's t-test, [c] Mann–Whitney U test for pairwise comparisons, as appropriate; p * Controls vs. All MS patients; p $ Controls vs. Established patients; p # Controls vs. Newly diagnosed (drug-naïve) patients.

3.1. Study Population

There was a total of 195 MS patients and 146 healthy controls. The patients' sample comprised 61.0% females and 39.0% males who were age-matched. The control population included 64.4% females and 35.6% males. The demographic and anthropometric features of both groups and the clinical phenotypes of the patients and the lifestyle habits (routine dressing, sunlight exposure) of both patients and controls are also indicated in Table 1.

With respect to the specific treatment at recruitment into the study (Table 1), 61 (or 31%) MS patients were newly diagnosed and not taking any disease-modifying drugs (newly diagnosed (drug-naïve)), while 134 (69%) had established disease and were on disease-modifying treatment (established MS). None of the patients were on steroid treatment within the period during and a month before recruitment into the study. In addition, 81 (42%) MS patients (comprising 60 RRMS, 18 SPMS, and 3 PPMS) and 33 (25%) healthy controls reported taking regular nutritional supplements, including multivitamins and vitamin D.

The patients were mostly young-to-middle-aged (median age 32 years), and the median age of MS diagnosis was <30 years; in addition, the median age at the onset of the symptoms was 27 years. Correspondingly, the duration of the disease at the time of recruitment was about 3 years.

3.2. Clinical Features

We also established that, in the patient group at the start of the study, 126 were in remission and 65 were in relapse. The major clinical phenotypes were: 166 (85.2%)

relapsing–remitting; 26 (13.3%) secondary progressive disease. Most patients (155 (79.5%)) had benign disease, with a median EDSS for all patients of just 1.5 (Table 1).

3.3. Serum 25(OH)D and DBP Levels

The median 25(OH)D levels (nmol/L) for both the healthy controls (28.4) and patients (27.3) were not different and fell within the vitamin D deficiency (close to severely deficient) category (Table 2). However, median serum DBP levels (μg/mL) were significantly lower in patients (163) than in controls (236; $p < 0.001$). Furthermore, the newly diagnosed (drug-naïve) MS group had lower 25(OH)D (18.9) and DBP (152) than the group with established disease (respectively 34.5, 183, $p < 0.001$). We confirmed that none of the subjects (patients and controls) had clinical or biochemical evidence of hepatic or renal disease and none of the females were on estrogen supplementation.

Table 2. Levels of 25(OH)D and DBP in the study population.

	Controls (n)	Patients with MS (n)			p [a]	p [b]	p [c]	p [d]
		All	Established MS	Newly Diagnosed (Drug-Naïve)				
n	146	195	134	61				
25(OH)D (nmol/L)	28.4 (18.0–50.4)	27.3 (17.0–50.8)	34.5 (22.1–52.7)	18.9 (13.0–35.9)	0.82	0.081	0.001	<0.001
DBP (μg/mL)	236 (152–288)	163 (123–241)	183 (123–241)	152 (117–235)	<0.001	<0.001	<0.001	0.401

n: the number of subjects. Numeric data are expressed as the medians (25th–75th percentile); p-values are generated via the Mann–Whitney U test for a pairwise comparisons. p [a] All MS patients vs. Controls; p [b] Established MS vs. Controls; p [c] New Diagnosed (drug-naïve) vs. Controls; p [d] Newly Diagnosed (drug-naïve) vs. Established MS patients.

We utilized logistic regression models to investigate potential interactions between serum levels of 25(OH)D and DBP and MS status in newly diagnosed (drug-naïve) patients who were not taking vitamin D supplementation. The results showed that:

i. There was an interaction between 25(OH)D and DBP on MS status in severely deficient (25(OH)D < 25 nmol/L) newly diagnosed (drug-naïve) patients.
ii. Low serum 25(OH)D with normal DBP had no significant impact on MS status (OR 1.02; 0.54–1.92; $p = 0.994$); low serum DBP showed a better, albeit non-significant, interaction (OR 1.56; 0.87–2.82; $p = 0.139$).
iii. The low 25(OH)D and DBP combination demonstrated a significant association with MS (OR 2.67; 1.35–5.29; $p = 0.005$).
iv. Serum 25(OH)D was within strata of DBP and appeared to be more strongly associated with MS when DBP levels were reduced (OR 1.71; 0.86–3.40, $p = 126$). This relationship is best demonstrated in the marginal probabilities diagram (Figure 1).

Marginal probabilities from logistic regression in newly diagnosed (drug-naïve) patients demonstrate the interaction between vitamin D binding protein and vitamin D deficiency based on the probability of it being an MS case. The highest probability of a case was in those with both vitamin D deficiency and a lower vitamin D binding protein level.

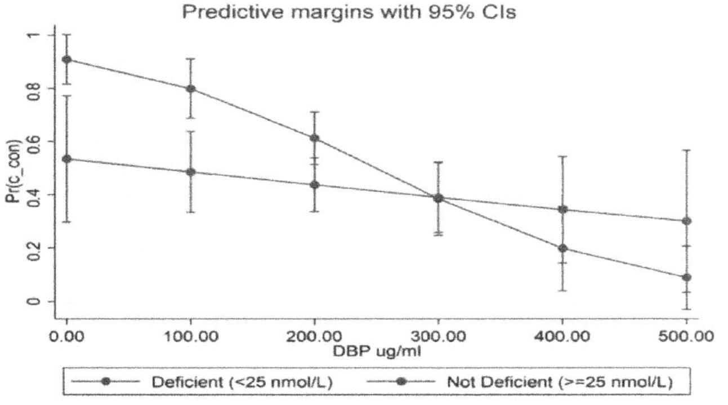

Figure 1. MS status predictive probability based on the logistic regression model.

3.4. Vitamin D and DBP Levels in Different MS Clinical Phenotypes

We evaluated the levels of 25(OH)D in the various clinical MS phenotypes with respect to the use or not use of vitamin D supplements in the patients' cohort (Table 3). The results indicated as expected, that all the patients, irrespective of disease phenotype (RRMS, SPMS) and use or not of vitamin D supplementation had 25(OH)D levels in the vitamin D deficiency category Note that the patients with Primary progressive MS (PPMS) were excluded from this analysis because of their relatively small sample size (n = 3). The results showed that independent of vitamin D supplementation status, the SPMS category had significantly higher 25(OH)D levels than controls (p = 0.046 & 0.056 respectively), and those with RRMS (p = 0.025 and 0.029, respectively). This is against the background that the DBP level was similar to that seen in RRMS but marginally significantly lower than the level in the healthy controls (p = 0.049) in subjects taking vitamin D supplements. Contrarywise, patients with SPMS and not on vitamin D supplementation had significantly higher DBP levels than those classified as RRMS. It is noteworthy that the patients with RRMS disease had similar 25(OH)D (p = 0.387) but significantly lower DBP levels ($p < 0.001$) when compared to the healthy control subjects.

Table 3. Serum levels of 25(OH)D and DBP in different MS phenotypes vs controls.

	Healthy Controls	Relapsing-Remitting	p [a]	Secondary Progressive	p [b]	p [c]
All Subjects (n)	146	166		26		
Subjects on Vitamin D supplement	33	60		16		
25(OH)D (nmol/L)	28.4 (18.0–50.4)	25.7 (16.0–47.8)	0.387	45.5 (22.0–59.2)	0.046	0.025
DBP (µg/mL)	236 (152–288)	162 (123–239)	<0.001	194 (127–246)	0.049	0.508
Subjects not on Vitamin D supplement	n = 112	n = 101		n = 10		
25(OH)D (nmol/L)	26.2 (16.7–35.0)	22.1 (13.0–36.3)	0.051	35.4 (19.4–64.7)	0.056	0.029
DBP (µg/mL)	246 (161–299)	159 (120–248)	<0.001	247 (222–271)	0.801	0.011

Values are expressed by Median (25th–75th percentile). p-values are generated using Mann–Whitney U test for pairwise comparisons [a] Relapsing-Remitting vs. Controls; [b] Secondary Progressive vs. Controls; [c] Relapsing-Remitting vs. Secondary Progressive. Note that Numbers may not add up to the total due to missing data.

Table 4 indicates the influence of disease activity (remission or relapse) on 25(OH)D and DBP levels in the sub-groups of patients who were newly diagnosed (drug-naïve)—these patients are likely to be free of confounding from the use of the disease-modifying therapies. The RRMS group on remission had similar 25(OH)D but lower DBP levels than the healthy control subjects. On the other hand, the group in relapse had significantly lower 25(OH)D and DBP levels when compared to controls (both $p < 0.05$). It is noteworthy that vitamin D level was lower in relapsing than in remitting disease, although the difference did not achieve statistical significance (Table 4).

Table 4. 25(OH)D and DBP levels in controls and newly diagnosed (drug-naïve) patients in relapse and remission.

	Controls	MS Relapse	p^a	MS Remission	p^b	p^c
N	146	36		25		
25(OH)D (nmol/L)	25.4 (16.3–33.5)	16.0 (12.5–26.5)	0.028	22.5 (14.1–47.2)	0.198	0.058
DBP (μg/mL)	247 (165–303)	155 (113–244)	<0.001	151 (118–233)	0.001	0.318

Values are expressed by the median (25th–75th percentile). *p*-values are generated using the Mann–Whitney U test for pairwise comparisons. p^a patients having relapse vs. controls, p^b patients having remission vs. controls; p^c patients having remission vs. relapse.

3.5. Correlation of Vitamin D and DBP with Disease Severity

We could not establish any significant correlation of EDSS with levels of 25(OH)D and/or DBP (Figure 2). However, serum 25(OH)D significantly positively correlated with the duration of the disease ($r = 0.242$; $p = 0.001$) (Figure 3). This could be ascribed to the frequent medical attention provided to patients with long-term MS with the consequence of increased awareness of the disease and the use of vitamin supplements apart from the specific medications.

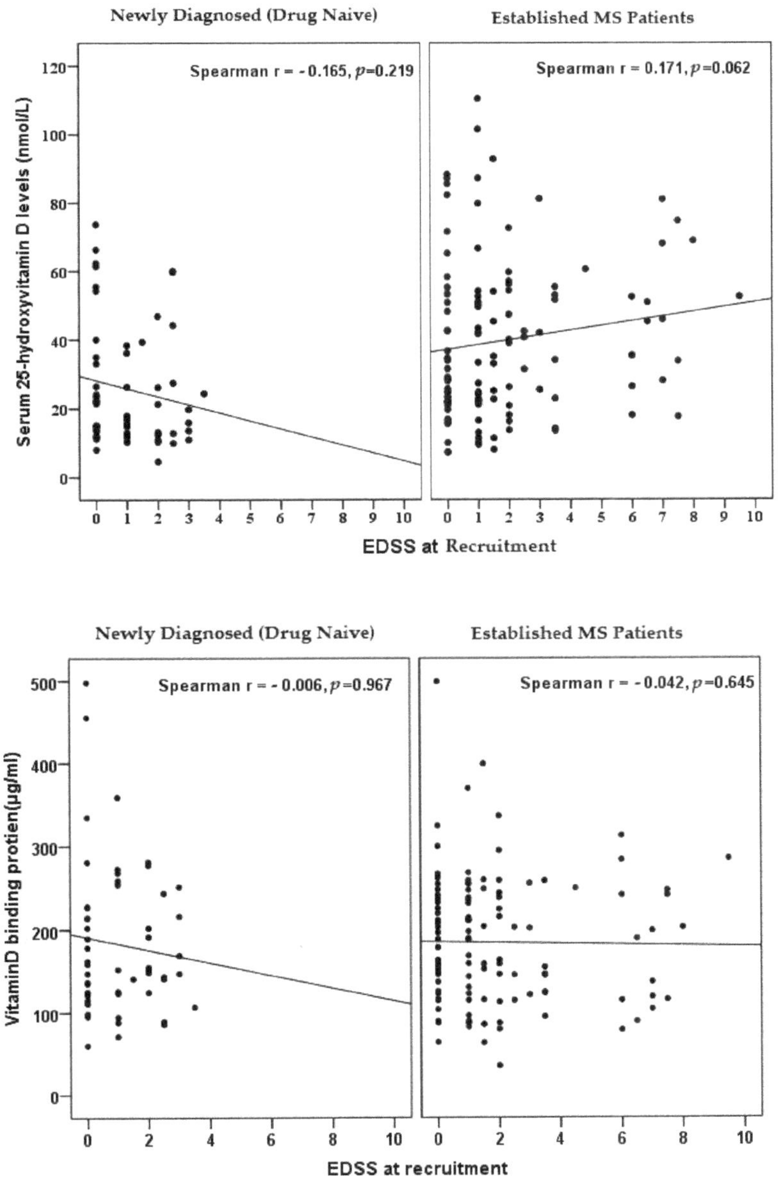

Figure 2. Correlation of vitamin D and Binding protein with EDSS. Serum level of 25(OH)D and VDBP are represented by black dots.

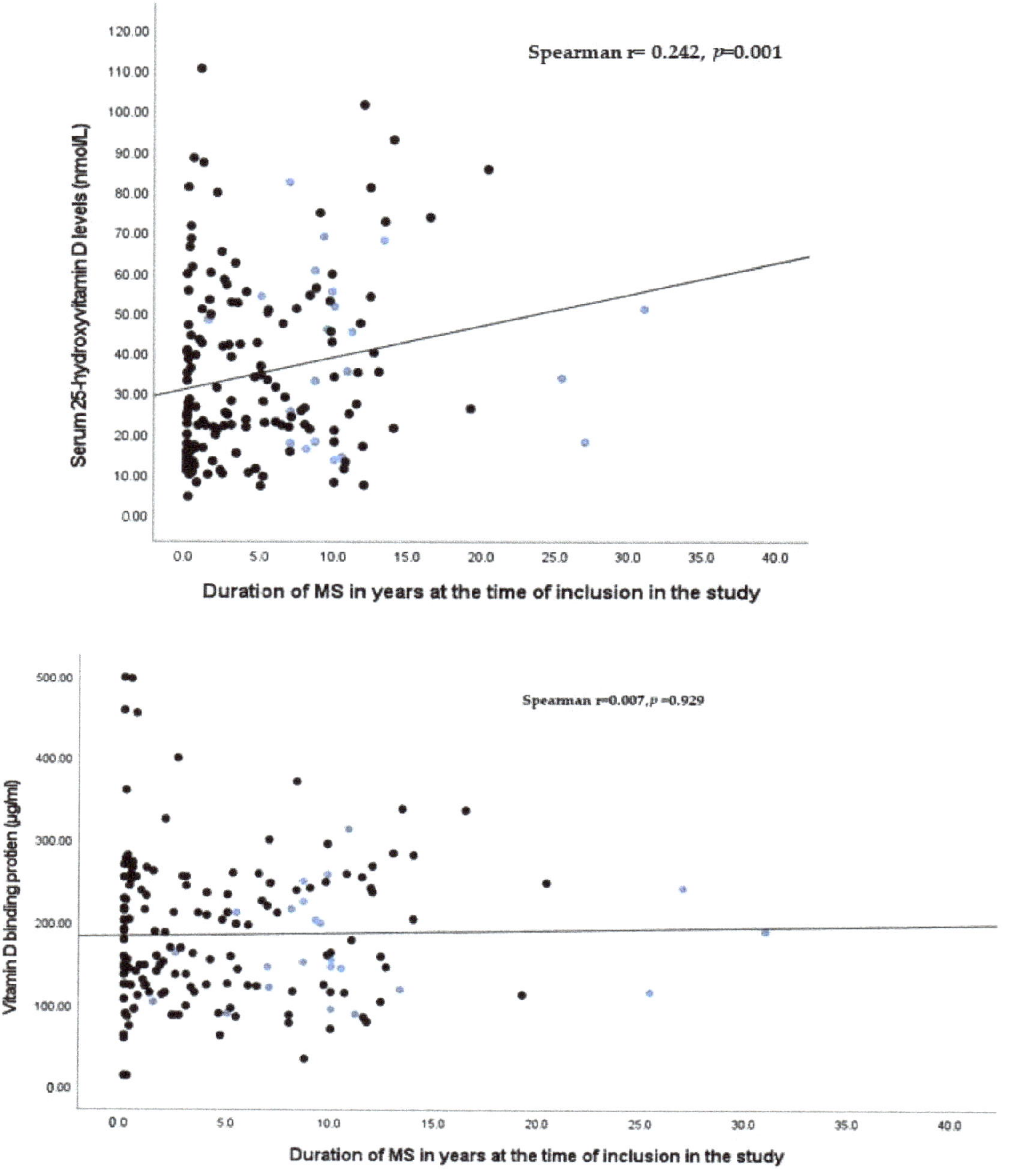

Figure 3. Correlation of 25(OH)D and DBP with duration of MS. Serum level of 25(OH)D and DBP are represented by dots. Black dots and blue dots represent RRMS and SPMS patients respectively.

4. Discussion

There is anecdotal evidence for increasing incidence and prevalence of MS in Kuwait. While two hospital-based studies reported an increased incidence in the last 2 decades [25–27], Some recent [27] and previous studies [28] suggested that Kuwait had emerged as a high-risk zone. This would be in keeping with the report by Alshubaili et al. [26], who reported that there was an apparent gradual increase in the incidence and prevalence of MS. If ultimately proven to be correct, the basis for this change might be unclear genetic, environmental, and sociocultural factors [24], perhaps ultimately acting via unexplored immunological mechanisms [25].

In this study, we have found that despite the year-round abundant sunlight in Kuwait, there is widespread vitamin D deficiency, as seen in both groups of healthy control subjects and patients with MS. The study also suggested an association of low 25(OH)D and low DBP with the development of MS, most clearly seen in the cohort of patients who were newly diagnosed (drug-naïve).

Other important findings from the study are: (i) the serum DBP level was inversely associated with MS prevalence; (ii) there is an interaction between circulating DBP and 25(OH)D and clinical MS; (iii). high DBP levels appear to attenuate the association of clinical MS status with concurrent vitamin D deficiency; (iv). there appears to be a > 2.5 X increase in the likelihood of MS development when serum levels of both DBP and 25(OH)D are low, and (v). the association of vitamin D with the development of MS appears strongest when DBP levels are low.

To the best of our knowledge, the synergism we have described between 25(OH)D and its carrier protein (DBP) in the development of MS and its clinical phenotypes is novel at least in our relatively homogeneous Arab population. The immediate deduction is that higher DBP protects against hypovitaminosis D-mediated MS risk. Whether this is a metabolic interaction or is precipitated by the other well-described independent effects of DBP is currently uncertain but deserves further study because of its potential therapeutic implications.

The plausibility of interaction between DBP and its ligand 25(OH)D to influence exposure risk is however not new. It has been observed previously in other biological systems including cancer [29].

At the molecular level, the uptake of the vitamin D-DBP complex in various organs and tissues [30]) is mediated by the plasma membrane receptor megalin. This is particularly important in the renal proximal tubules for resorption of DBP-25(OH)D from the glomerular filtrate and subsequently conversion of 25(OH)D to the bioactive form, 1,25(OH)$_2$D. It has been demonstrated that megalin knockout mice are incapable of resorbing 25(OH)D–DBP complex from urine resulting in loss of 25(OH)D and profoundly low circulating 25(OH)D and 1,25(OH)$_2$D levels. It is therefore possible that higher circulating DBP promotes greater renal tubule DBP resorption and higher bioavailable 1,25(OH)$_2$D to target tissue and mitigate a putative MS risk related to low circulating 25(OH)D levels. The presence of megalin in the brain will provide some evidence for the role of DBP in increasing the neural bioavailability of vitamin D and subsequent protection from demyelination.

In the current study, the plasma level of DBP was significantly lower in patients than in the healthy controls and lowest in the RRMS-relapse group. This is consistent with previous observations [12–14]. Some other studies reported no change in circulating DBP in relapsing MS during remission and relapse [16] and during remission [12] as compared to healthy controls, while at least one study [18] reported higher DBP levels in RRMS. We believe that these discrepant observations may be due to, among other variables, issues with sample population size and heterogenicity (our population was demographically and socioculturally relatively homogenous), and DBP analytical methods [31].

Lower DBP and 25(OH)D levels, as observed in our study, might indicate that free and bioavailable vitamin D levels are also low, although we did not specifically calculate a free vitamin D index [13]. Both those latter studies indeed did demonstrate that low levels of DBP influence MS risk, even at the pre-symptomatic phase. It seems reasonable to suggest that the combination of low DBP and low 25(OH)D will attempt to preserve the free vitamin D index for physiologic ubiquitous vitamin D function. However, with intensified tissue inflammation and/or lack of disease modification with drugs [as seen in the newly diagnosed (drug-naïve) patients or those with MS relapse], continuing reductions in 25(OH)D and DBP (possibly as part of the acute phase response), will upset the vitamin D/DBP equilibrium, with changes in the free vitamin D index and tendency towards the clinical expression of MS. This is however speculative and subject to confirmation. What is non-controversial is that DBP potentially influences the immune system [32] and may have a favorable impact on immune regulation independent of, and in association with the immunoregulatory functions of vitamin D [measured as 25(OH)D] in the pathogenesis of MS. Certainly, the genetic polymorphism of DBP itself (not explored on this study) may also play a role in this interaction.

There have been a few reports on the potential role of brain DBP (as measured by CSF analyses) in the pathogenesis of MS. Some CSF proteomic studies found lower levels of DBP in MS [10,11] in relapse and in comparison, to controls with other neurologic disorders. Other recent studies reported lower DBP CSF levels during the acute phase of demyelinating attacks in MS patients [33], which appears to mirror observations in the peripheral circulation [12,14]. However, these observations are not consistent [16].

This study could not support reports from some other studies [17] of significant associations between vitamin D levels and measures of disability (EDSS). We believe that such associations might be possible with larger subject numbers and will form the basis of our continuing studies. DBP levels are significantly influenced by genetic polymorphism and being acute phase proteins might also be influenced by medications and concomitant physical illnesses. All these, with vitamin D nutriture (that can be influenced by diet in the chronically ill, even in a relatively high-income population) will affect the links between vitamin D, DBP, and MS prevalence and phenotypes. Our study did not specifically evaluate the interplay of all these issues and should possibly be considered a pilot. There are ongoing efforts to investigate in more detail these mechanistic issues. Nonetheless, this study had several strengths, including a larger sample size, the inclusion of both drug-naïve patients and those who are on disease-modifying therapies, and analyzing a patient population that included different MS phenotypes and individuals in remission and/or relapse. One important limitation however is the relatively small number of SPMS and the lack of measurement of free vitamin D and indicators of DBP genetic polymorphism.

5. Conclusions

The study has demonstrated that 25(OH)D deficiency is common in both Kuwaiti patients with MS and apparently healthy controls, probably because of a wide variety of social, environmental, and genetic factors. We have suggested an association between low levels of 25(OH)D and DBP and the presence of MS, which appeared strongest in the drug-naïve MS patients, whose biochemical measurements could not have been confounded by medications. Additionally, MS patients in relapse had relatively low 25(OH)D and DBP levels. We could not establish associations between serum 25(OH)D and DBP concentrations and disability in MS. We suggest that the homeostatic regulation of the relationship between circulating 25(OH)D level and its binding protein, DBP, may be impaired in the pathogenetic process involved in the onset of MS and its progression into distinct clinical phenotypes. This may have been triggered by yet unknown genetic and environmental factors and which will form the basis for our further studies.

Author Contributions: Conceptualization, S.A.-S. and A.A.; methodology, S.A.-S. and A.A.; software, A.C.; validation, S.A.-S., A.A. and A.C.; formal analysis, A.C. and S.D; investigation, S.A.-S., A.C. and M.G.H.; resources, S.A.-S. and M.G.H.; data curation, A.C.; writing—original draft preparation, S.A.-S., A.C., S.A.-S., A.A., A.C. and S.D. writing—review and editing, S.A.-S. and A.A.; visualization, S.A.-S. and A.A.; supervision, S.A.-S. and A.A.; project administration, S.A.-S. and A.C.; funding acquisition, S.A.-S. and A.A. All authors have read and agreed to the published version of the manuscript.

Funding: This research was funded by Kuwait University Research Administration; grant number MM 03/09 and the APC was funded by Suhail Al-Shammri.

Institutional Review Board Statement: The study was conducted in accordance with the Declaration of Helsinki, and approved by the Ethical Committee Faculty of Medicine, Kuwait University (IORG0007925 and 1 July 2010).

Informed Consent Statement: Informed consent was obtained from all subjects involved in the study.

Data Availability Statement: The database generated and/or analyzed during the current study are not publicly available due to ethical considerations but are available from the corresponding author on reasonable request.

Acknowledgments: The authors thank Edison Gomes for his contribution in statistical analysis.

Conflicts of Interest: The authors declare no conflict of interest. The funder of the project (Research Administration, Kuwait university) had no role in the design of the study; in the collection, analyses, or interpretation of data; in the writing of the manuscript; or in the decision to publish the results, however funder of APC is the corresponding author.

References

1. Lassmann, H.; Brück, W.; Lucchinetti, C.F. The Immunopathology of Multiple Sclerosis: An Overview. *Brain Pathol.* **2007**, *17*, 210–218. [CrossRef] [PubMed]
2. Noseworthy, J.H.; Lucchinetti, C.; Rodriguez, M.; Weinshinker, B.G. Medical progress: Multiple sclerosis. *N. Engl. J. Med.* **2000**, *343*, 938–952. [CrossRef]
3. Simon, K.C.; Munger, K.I.; Yang, X.; Ascherio, A. Polymorphism in vitamin D metabolism related genes and risk of multiple sclerosis. *Mult. Scler.* **2010**, *16*, 133–138. [CrossRef]
4. Munger, K.L.; Levin, L.I.; Hollis, B.W.; Howard, N.S.; Ascherio, A. Serum 25-hydroxyvitamin D levels and risk of multiple sclerosis. *JAMA* **2006**, *296*, 2832–2838. [CrossRef] [PubMed]
5. Simpson Jr., S.; Taylor, B.; Blizzard, L.; Ponsonby, A.L.; Pittas, F.; Tremlett, H.; Dwyer, T.; Gies, P.; van der Mei, I. Higher 25-hydroxyvitamin D is associated with lower relapse risk in multiple sclerosis. *Ann. Neurol.* **2010**, *68*, 193–203. [CrossRef]
6. Van der Mei, I.A.; Ponsonby, A.L.; Dwyer, T.; Blizzard, L.; Taylor, B.V.; Kilpatrick, T.; Butzkueven, H.; McMichael, A.J. Vitamin D levels in people with multiple sclerosis and community controls in Tasmania, Australia. *J. Neurol.* **2007**, *254*, 581–590. [CrossRef]
7. Mowry, E.M.; Waubant, E.; McCulloch, C.E.; Okuda, D.T.; Evangelista, A.A.; Lincoln, R.R.; Gourraud, P.A.; Brenneman, D.; Owen, M.C.; Qualley, P.; et al. Vitamin D status predicts new brain magnetic resonance imaging activity in multiple sclerosis. *Ann. Neurol.* **2012**, *72*, 234–240. [CrossRef] [PubMed]
8. White, P.; Cook, N. The multifunctional properties and characteristics of vitamin D-binding protein. *Trends Endocrinol. Metab.* **2000**, *11*, 320–327. [CrossRef]
9. Chun, R.F. New perspectives on vitamin D binding protein. *Cell Biochem. Func.* **2012**, *30*, 445–456. [CrossRef]
10. Qin, Z.; Qin, Y.; Liu, S. Alteration of DBP levels in CSF of patients with MS by proteomics analysis. *Cell. Mol. Neurobiol.* **2009**, *29*, 203–210. [CrossRef] [PubMed]
11. Liu, S.; Bai, S.; Qin, Z.; Yang, Y.; Cui, Y.; Qin, Y. Quantitative proteomic analysis of the cerebrospinal fluid of patients with multiple sclerosis. *J. Cell. Mol. Med.* **2009**, *13*, 1586–1603. [CrossRef]
12. Niino, M.; Fukazawa, T.; Miyazaki, Y.; Minami, N.; Tashiro, J.; Amino, I.; Nonaka, T.; Fujiki, N.; Doi, S.; Kikuchi, S. Association of Vitamin D levels in Japanese patients with Multiple Sclerosis. *Clin. Exp. Neuroimmunol.* **2013**, *4*, 193–200. [CrossRef]
13. Grut, V.; Biström, M.; Salzer, J.; Stridh, P.; Lindam, A.; Alonso-Magdalena, L.; Andersen, O.; Jons, D.; Gunnarsson, M.; Vrethem, M.; et al. Free vitamin D3 index and vitamin D-binding protein in multiple sclerosis: A presymptomatic case-control study. *Eur. J. Neurol.* **2022**, *29*, 2335–2342. [CrossRef] [PubMed]
14. Maghbooli, Z.; Omidifar, A.; Varzandi, T.; Salehnezhad, T.; Sahraian, M.A. Reduction in circulating vitamin D binding protein in patients with multiple sclerosis. *BMC Neurol.* **2021**, *21*, 168. [CrossRef] [PubMed]
15. Yang, M.; Qin, Z.; Zhu, Y.; Li, Y.; Qin, Y.; Jing, Y.; Liu, S. Vitamin D-binding protein in cerebrospinal fluid is associated with multiple sclerosis progression. *Mol. Neurobiol.* **2013**, *47*, 946–956. [CrossRef] [PubMed]
16. Smolders, J.; Peelen, E.; Thewissen, M.; Menheere, P.; Damoiseaux, J.; Hupperts, R. Circulating vitamin D binding protein levels are not associated with relapses or with vitamin D status in multiple sclerosis. *Mult. Scler.* **2014**, *20*, 433–437. [CrossRef]
17. Aktürk, T.; Turan, Y.; Tanik, N.; Karadağ, M.E.; Sacmaci, H.; Inan, L.E. Vitamin D, vitamin D binding protein, vitamin D receptor levels and cardiac dysautonomia in patients with multiple sclerosis: A cross-sectional study. *Arq Neuropsiquiatr.* **2019**, *77*, 848–854. [CrossRef]
18. Rinaldi, A.O.; Sanseverino, I.; Purificato, C.; Cortese, A.; Mechelli, R.; Francisci, S.; Salvetti, M.; Millefiorini, E.; Gessani, S.; Gauzzi, M.C. Increased circulating levels of vitamin D binding protein in MS patients. *Toxins* **2015**, *7*, 129–137. [CrossRef] [PubMed]
19. Wacholder, S.; Silverman, D.T.; McLaughlin, J.K.; Mandel, J.S. Selection of controls in case-control studies. III. Design options. *Am. J. Epidemiol.* **1992**, *135*, 1042–1050. [CrossRef]
20. Polman, C.H.; Reingold, S.C.; Banwell, B.; Clanet, M.; Cohen, J.A.; Filippi, M.; Fujihara, K.; Havrdova, E.; Hutchinson, M.; Kappos, L.; et al. Diagnostic criteria for multiple sclerosis. Revision to the McDonald criteria. *Ann. Neurol.* **2011**, *69*, 292–302. [CrossRef]

21. Kurtzke, J.F. Rating neurologic impairment in multiple sclerosis an expanded disability status scale (EDSS). *Neurology* **1983**, *33*, 1444–1452. [CrossRef]
22. Holick, M.F.; Binkley, N.C.; Bischoff-Ferrari, H.A.; Gordon, C.M.; Hanley, D.A.; Heaney, R.P.; Murad, M.H.; Weaver, C.M. Evaluation, treatment, and prevention of vitamin D deficiency: An Endocrine Society Clinical Practice Guideline. *J. Clin. Endocrinol. Metab.* **2011**, *96*, 1911–1930. [CrossRef]
23. Braegger, C.; Campoy, C.; Colomb, V.; Decsi, T.; Domellof, M.; Fewtrell, M.; Hojsak, I.; Mihatsch, W.; Molgaard, C.; Shamir, R.; et al. Vitamin D in the healthy european paediatric population. *J. Pediatr. Gastroenterol. Nutr.* **2013**, *56*, 692–701. [CrossRef]
24. Al-Shammri, S.N.; Hanna, M.G.; Chattopadhyay, A.; Akanji, A.O. Sociocultural and Demographic Risk Factors for the Development of Multiple Sclerosis in Kuwait: A Case-Control Study. *PLoS ONE* **2015**, *10*, e0132106. [CrossRef]
25. Al-Din, A.S.N.; Khogali, M.; Poser, C.M.; Al-Nassar, K.E.; Shakir, R.; Hussain, J.; Behbahani, K.; Chadha, G. Epidemiology of multiple sclerosis in Arabs in Kuwait—a comparative study between Kuwaitis and Palestinians. *J. Neurol. Sci.* **1990**, *100*, 137–141. [CrossRef]
26. Al Shubaili, A.F.; Alramzy, K.; Ayyad, Y.M.; Gerish, Y. Epidemiology of multiple sclerosis in Kuwait: New trends in incidence and prevalence. *Eur. Neurol.* **2005**, *53*, 125–131. [CrossRef] [PubMed]
27. Alroughani, R.; Ahmed, S.; Behbahani, R.; Khan, R.; Thussu, A.; Alexander, K.J.; Ashkanani, A.; Nagarajan, V.; Al-Hashel, J. Inreasing prevalence and incidence rates of multiple sclerosis in Kuwait. *Mult. Scler.* **2014**, *20*, 543–547. [CrossRef]
28. Kurtzke, J.F. MS epidemiology worldwide. One view of current status. *Acta Neurol. Scand.* **1995**, *91* (Suppl. 161), 23–33. [CrossRef] [PubMed]
29. Weinstein, S.J.; Stolzenberg-Solomon, R.Z.; Kopp, W.; Rager, H.; Virtamo, J.; Albanes, D. Impact of Circulating Vitamin D Binding Protein Levels on the Association between 25-Hydroxyvitamin D and Pancreatic Cancer Risk: A Nested Case–Control Study. *Cancer Res.* **2012**, *72*, 1190–1198. [CrossRef] [PubMed]
30. Nykjaer, A.; Fyfe, J.C.; Kozyraki, R.; Leheste, J.R.; Jacobsen, C.; Nielsen, M.S.; Verroust, P.J.; Aminoff, M.; de la Chapelle, A.; Moestrup, S.K.; et al. Cubilin dysfunction causes abnormal metabolism of the steroid hormone 25(OH) vitamin D (3). *Proc. Natl. Acad. Sci. USA* **2001**, *98*, 13895–13900. [CrossRef]
31. Gauzzi, M.C. Vitamin D-binding protein and multiple sclerosis: Evidence, controversies, and needs. *Mult. Scler.* **2018**, *24*, 1526–1535. [CrossRef] [PubMed]
32. Delanghe, J.R.; Speeckaert, R.; Speeckaert, M.M. Behind the scenes of vitamin D binding protein: More than vitamin D binding. *Best Pract. Res. Clin. Endocrinol. Metab.* **2015**, *29*, 773–786. [CrossRef] [PubMed]
33. Lehmensiek, V.; Sussmuth, S.D.; Tauscher, G.; Brettschneider, J.; Felk, S.; Gillardon, F.; Tumani, H. Cerebrospinal fluid proteome profile in multiple sclerosis. *Mult. Scler.* **2007**, *13*, 840–849. [CrossRef] [PubMed]

biomedicines

MDPI

Article

Oxidative Stress Markers and Heat Shock Proteins in Non-Obese Women with Polycystic Ovary Syndrome Are Not Elevated and Show No Correlation with Vitamin D

Manjula Nandakumar [1], Thozhukat Sathyapalan [2], Alexandra E. Butler [1,*,†] and Stephen L. Atkin [1,†]

1 Research Department, Royal College of Surgeons in Ireland Bahrain, Busaiteen, Adliya 15503, Bahrain; mnandakumar@rcsi.com (M.N.); satkin@rcsi.com (S.L.A.)

2 Academic Endocrinology, Diabetes and Metabolism, Hull York Medical School, Hull HU6 7RU, UK; thozhukat.sathyapalan@hyms.ac.uk

* Correspondence: aeb91011@gmail.com or abutler@rcsi.com; Tel.: +973-32360292

† These authors contributed equally to this work.

Abstract: Introduction. Oxidative stress (OS) is recognized in the pathophysiology of polycystic ovary syndrome (PCOS). OS results in intracellular reactive oxygen species generation, causing oxidative protein damage that is protected by heat shock proteins (HSPs). Vitamin D is thought to reduce and protect against OS; therefore, OS, HSP, and vitamin D levels may be associated with PCOS. However, their expression in PCOS without underlying inflammation is unknown. Methods. In this exploratory study, the plasma levels of 7 OS proteins and 10 HSPs that are affected by the OS process were measured using Slow Off-rate Modified Aptamer (SOMA)-scan plasma protein measurements in non-obese, non-insulin resistant women with PCOS (n = 24) without systemic inflammation and control (n = 24) women; the cohorts were matched for weight and age. The OS proteins and HSPs were correlated with 25-hydroxy vitamin D_3 (25(OH)D_3) and the active form, 1,25-dihydroxyvitamin D_3 (1,25(OH)$_2D_3$), as measured by isotope-dilution liquid chromatography tandem mass spectrometry. Results. The PCOS women versus the controls had comparable insulin resistance and systemic inflammation (C-reactive protein 2.0 mg/L vs. 2.3 mg/L, $p > 0.05$), but higher free androgen index and anti-mullerian hormone levels. Among the OS proteins, only esterase D (ESD; $p < 0.01$) was elevated in PCOS and the HSPs did not differ between the PCOS and control women. There was no correlation of 25(OH)D_3 or 1,25(OH)$_2D_3$ with any of the proteins. Conclusions. In a PCOS population that was non-obese and without insulin resistance and systemic inflammation, only ESD was elevated in PCOS, whilst the other OS proteins and HSPs were not elevated. Further, none of the OS proteins or HSPs were correlated with either 25(OH)D_3 or 1,25(OH)$_2D_3$ in either cohort of women or when both cohorts were combined, indicating that the OS and HSP responses were largely absent and not affected by vitamin D in a non-obese PCOS population.

Keywords: polycystic ovary syndrome; oxidative stress; heat shock proteins; vitamin D_3

1. Introduction

Affecting 5–10% of women, polycystic ovary syndrome (PCOS) is the most prevalent endocrine abnormality in premenopausal women [1]. PCOS is a diagnosis of exclusion that can be determined by recognized international criteria [2] and it is estimated that 70% of women with PCOS remain undiagnosed [3]. It is well recognized that polycystic ovary syndrome (PCOS) is a metabolic disease with an increased prevalence of type 2 diabetes, hypertension, dyslipidemia, and enhanced cardiovascular risk factors in affected women [4]. The pathophysiological mechanisms underlying these features found in PCOS are still unclear, though oxidative stress (OS), perhaps primarily due to obesity, has been implicated [5,6]; however, other mechanisms involving OS mediated through hyperandrogenemia, vitamin D levels, and insulin resistance have also been reported [7,8], together with other cellular biomarkers such as the dysregulation of heat shock proteins [9], coagulation markers [10], and the complement pathway proteins, through a combination of obesity and insulin resistance as the underlying mediators [11–14].

Whilst obesity is common in PCOS, it is likely not a primary cause, as is reflected in the high prevalence of PCOS among non-obese populations [15]. However, obesity contributes to the exacerbation of its underlying pathophysiology, especially cardiovascular risk factors such as insulin resistance, glucose intolerance, and dyslipidemia. Weight loss decreases these cardiovascular risk markers, as is particularly noted following bariatric surgery [16]; however, no medical intervention has been shown to decrease cardiovascular events or delay the onset of diabetes [17]. Chronic systemic inflammation is seen in obesity, which is reflected in the enhanced serum levels of inflammatory cytokines and lymphocyte function changes [18–20]. High-sensitivity C-reactive protein (hs-CRP) is a measure of inflammation and women with PCOS have significantly increased hs-CRP concentrations [21]. It has been recognized that CRP elevation is independently related to insulin resistance and may be related to coronary heart disease and cardiovascular events in PCOS [22]. CRP has been shown to be a strong predictor of cardiovascular events in women [23].

PCOS is associated with insulin resistance, an increased risk of diabetes, non-alcoholic fatty liver disease (NAFLD), and increased cardiovascular (CV) risk, independent of obesity [24,25]. Insulin resistance results from a defect in insulin action, including insulin-mediated glucose transport and its signaling pathway [26]. Hyperinsulinemia results in an increased risk of type 2 diabetes in PCOS [27]. In addition, hyperinsulinemia leads to dyslipidaemia, which is seen as a common feature in PCOS and contributes to its documented cardiometabolic abnormalities [28].

Increased OS is caused by an imbalance between oxidants and antioxidants that then results in the formation of intracellular reactive oxygen species (ROS). The circulating markers of OS have been reported by some as being abnormal in PCOS [29], though these reports have been conflicting [5]. The generation of ROS leads to protein oxidation that may affect many biological parameters, ultimately resulting, for example, in vascular injury [30]. However, OS protein damage is limited by the protective HSP response [31]. HSPs function as molecular chaperones that prevent protein misfolding in order to prevent intracellular dysregulation [32], and they are constitutively expressed at low levels but become elevated with a stress stimulus such as OS [33] (Figure 1). Impaired proteins that are misfolded could result in cellular dysregulation and are channeled by HSPs to the ubiquitin proteasome system (UPS) to effect their degradation [34]; following ubiquination, proteolysis occurs via the 26S proteasome [35]. HSPs can be seen to be involved in the regulation and functional facilitation of the critical enzymes in inflammation, apoptosis, metabolism, and cell signaling [36,37]. HSPs are becoming increasingly recognized as having an important role in the pathophysiology of PCOS [9], acting as protection against PCOS-associated increased OS, inflammation, and insulin resistance [9].

Vitamin D deficiency is highly prevalent in women with PCOS, with 67–85% exhibiting a severe deficiency. Vitamin D deficiency is associated with the PCOS pathognomonic features of obesity, increased insulin resistance, and elevated levels of testosterone [38,39]; however, whilst vitamin D deficiency may not exacerbate these features in the setting of PCOS [40], vitamin D replacement has been suggested to benefit both insulin resistance and steroidogenesis in obese women with PCOS [41,42]. The vitamin D deficiency reported to be associated with insulin resistance is considered to be obesity dependent [43] and, indeed, vitamin D deficiency increases with obesity as the vitamin D is sequestered in adipose tissue. Vitamin D has been suggested to be a key controller of systemic inflammation, oxidative stress, and mitochondrial function and is a potent antioxidant in its own right [44] Vitamin D_3 (cholecalciferol) is produced endogenously in the skin as a consequence of ultraviolet B radiation (UV-B), acting on 7-dehydrocholesterol that is hydroxylated at position 25 to 25-hydroxy vitamin D_3 ($25(OH)D_3$). $25(OH)D_3$ is transported to the kidney, where 1-alpha hydroxylase converts it into the active form, $1,25(OH)_2D_3$ [45].

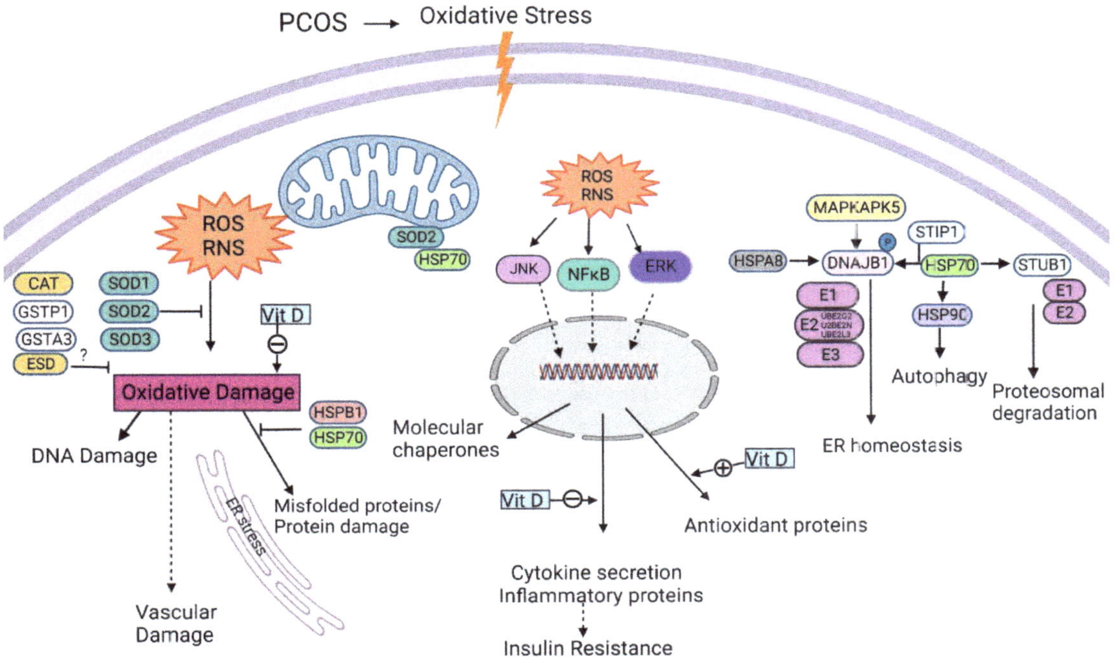

Figure 1. Schematic illustration to show the interactions of vitamin D, heat shock proteins, and oxidative-stress-related proteins. In PCOS-associated oxidative stress, the ROS/RNS produced are linked to several pathways (examples being the NFkB, JNK, and ERK signaling pathways) which are involved in regulating pro-oxidant genes which lead to oxidative damage and antioxidant genes which protect from oxidative damage. Oxidative stress results in protein misfolding and oxidative damage of proteins, thereby causing ER stress. Heat shock proteins work to reduce the ER stress and maintain ER homeostasis. Many proteins which play a role as antioxidants are produced during oxidative stress which, by themselves or together with HSPs, help in reducing oxidative damage. Vitamin D is reported to reduce oxidative-stress-related proteins such as inflammatory proteins/secreted cytokines and to upregulate antioxidant enzymes/proteins. HSP90 alpha (HSP90AA1, HSP90AB1, HSP90 beta, and HSP90 dimer); HSPA8, heat shock cognate 71 kDa protein; DNAJB1, DnaJ homolog subfamily B member 1; MAPKAPK5, MAP kinase-activated protein kinase 5; STIP1, stress-induced phosphoprotein 1; E1, ubiquitin-activating enzyme; E2, ubiquitin-conjugating enzyme 2; UBE2L3, ubiquitin-conjugating enzyme E2 L3; UBE2N, ubiquitin-conjugating enzyme E2 N; E3, ubiquitin ligases; STUB1, STIP1 Homology And U-Box Containing Protein 1; SOD1, Superoxide dismutase1; SOD2, Superoxide dismutase2; SOD3, Superoxide dismutase3; CAT, Catalase; ESD, Esterase D; GSTP1, Glutathione S-transferase P1; GSTA3, Glutathione S-transferase A3; JNK, c-Jun N-terminal kinases; NFkB, Nuclear factor-kappa B; ERK, extracellular signal-regulated kinases; Vit D, Vitamin D; HSP90 alpha (HSP90AA1, HSP90AB1, HSP90 beta, HSP90 dimer); HSPA8, heat shock cognate 71 kDa protein; DNAJB1, DnaJ homolog subfamily B member 1; MAPKAPK5, MAP kinase-activated protein kinase 5; STIP1, stress-induced phosphoprotein 1; E1, ubiquitin-activating enzyme; E2, ubiquitin-conjugating enzyme 2; UBE2L3, ubiquitin-conjugating enzyme E2 L3; UBE2N, ubiquitin-conjugating enzyme E2 N; E3, ubiquitin ligases; STUB1, STIP1 Homology And U-Box Containing Protein 1; SOD1, Superoxide dismutase1; SOD2, Superoxide dismutase2; SOD3, Superoxide dismutase3; CAT, Catalase; ESD, Esterase D; GSTP1, Glutathione S-transferase P1; and GSTA3, Glutathione S-transferase A3.

Obesity, inflammation, and insulin resistance, with accompanying OS, are highly correlated with PCOS; thus, statistical regression adjustment for BMI/insulin resistance/inflammation is misleading, as it will over-adjust the effects of PCOS. Therefore, this study was performed on a cohort of women with PCOS who were non-obese versus control women; the cohorts were matched for BMI, insulin resistance, and systemic inflammation to determine the association of circulatory OS proteins and HSPs with PCOS and whether there was evidence of modulation by vitamin D metabolites.

2. Materials and Methods

The approval for this study was granted by The Yorkshire and The Humber NRES ethical committee, UK (February 2003; study approval number 02/03/043).

Seven plasma OS protein levels and ten plasma HSP levels were determined in 24 women with PCOS and 24 control women, all of whom were patients attending the Hull IVF clinic [45]. The ages and body mass indexes (BMI) of the control women were matched to those of the women with PCOS. For each of the two cohorts, the control and PCOS women, the demographic data are detailed are Table 1 [45]. The Rotterdam consensus criteria were used for diagnosis, following the exclusion of confounding conditions such as non-classical 21-hydroxylase deficiency, Cushing's disease, hyperprolactinemia, or an androgen-secreting tumor. The specific Rotterdam criteria for the diagnosis of PCOS include oligomenorrhea or amenorrhea; biochemical hyperandrogenemia or clinical hyperandrogenemia with a Ferriman–Gallwey score of >8 plus a free androgen index of >4; and polycystic ovaries, as determined by transvaginal ultrasound [46]. The study subjects had no other conditions/illnesses and the study's inclusion criteria required them to be medication-free (including over-the-counter antioxidants) for at least the 9 months prior to enrollment in the study. The procedures performed on the women in this study were aligned with the Helsinki declaration and its later amendments or comparable ethical standards.

Table 1. Demographics, baseline, hormonal, and metabolic parameters of the PCOS subjects and controls (mean ± SD). All parameters did not differ other than those marked ** = $p < 0.01$.

	Control (n = 24)	PCOS (n = 24)
Age (years)	32.5 ± 4.1	31 ± 6.4
BMI (kg/m^2)	24.8 ± 1.1	25.9 ± 1.8
Fasting glucose (mmol/L)	4.9 ± 0.4	4.7 ± 0.8
HbA1C (mmol/mol)	30.9 ± 6.5	31.8 ± 3.0
HOMA-IR	1.8 ± 1.0	1.9 ± 1.6
SHBG (nmol/L)	104 ± 80	72 ± 62
Free androgen index (FAI)	1.3 ± 0.5	4.1 ± 2.9 **
CRP (mg L^{-1})	2.3 ± 2.3	2.7 ± 2.5
AMH (ng/mL)	24 ± 13	57 ± 14 **
25 hydroxy vitamin D_3 (nmol/l)	46.2± 23.5	54.0 ± 27.4
1,25 Dihydroxy vitamin D_3 (ng/mL)	0.03 ± 0.02	0.04 ± 0.2

BMI—Body Mass Index; HbA1c—glycated hemoglobin; HOMA-IR—Homeostasis model of assessment—insulin resistance; CRP—C reactive protein; SHBG—sex hormone binding globulin; and AMH—Anti-Müllerian hormone.

Seasonal fluctuations in vitamin D metabolite levels are well recognized and may contribute to their variability. Nine minutes of daily sunlight exposure would be sufficient for vitamin D sufficiency in northern England; therefore, to ensure the targets were met and to reduce potential variability, the subjects were recruited from March to September [47]. Vitamin D supplement consumption was an exclusion criterion of the study.

Bloods drawn from the women in the fasting state were centrifuged at 3500× g for 15 min, then aliquoted and frozen (−80 °C) in preparation for analysis. The aliquots of blood were analyzed for the following: insulin, sex hormone binding globulin (SHBG) (DPC Immulite 200 analyser, Euro/DPC, Llanberis, UK), and plasma glucose (Synchron LX20 analyser, Beckman-Coulter, High Wycombe, UK). To calculate the free androgen index (FAI), the total testosterone was divided by the SHBG, and the result was multiplied by 100. To calculate the insulin resistance (IR), the homeostasis model assessment (HOMA-IR) was employed. To quantify the serum testosterone, isotope-dilution liquid chromatography tandem mass spectrometry (LC-MS/MS) methodology was utilized [45]. Likewise, to quantify the serum vitamin D levels, isotope-dilution liquid chromatography tandem mass spectrometry (LC-MS/MS) was utilized. Vitamin D metabolites (25(OH)D_3 and 1,25$(OH)_2D_3$), together with labeled internal standards (d6-25(OH)D_3 and d6-1,25$(OH)_2D_3$), were extracted simultaneously from 250 μL of serum with the use of supportive liquid–liquid extraction and Diels–Alder derivatization being undertaken prior to performing an LC-MS/MS analysis. To obtain chromatographic separations, the following was employed: a Hypersil Gold C18 column (150 × 2.1 mm; 1.9 μ) at a flow rate of 0.2 mL/min, set to operate in the Electrospray Ionisation (ESI) positive mode, and an analysis performed using the reaction monitoring (MRM) method. The limit of quantification (LOQ) was determined to be as follows: 0.5 ng/mL for 25(OH)D_3 and 10 pg/mL for 1,25$(OH)_2D_3$, with a 25(OH)D_3 cutoff of less than 20 ng/L (50 nmol/L) to define deficiency.

The circulatory levels of plasma proteins were measured using Slow Off-rate Modified Aptamer (SOMA)-scan plasma protein measurement technology (Somalogic, Boulder, CO, USA), as has been described in detail in prior reports [48]. Raw intensity normalization and hybridization, together with median signal and calibration signal determination, were based upon the referent standard samples that are incorporated into each plate, as has been previously detailed [49].

Measurements of the plasma levels of the following seven OS proteins were undertaken: Superoxide dismutase (SOD1); Glutathione S-transferase P (GSTP1); Glutathione S-transferase A3 (GSTA3); Esterase D (ESD); Catalase (CAT); Superoxide dismutase mitochondrial (SOD2); and Extracellular superoxide dismutase (SOD3). In addition, we measured 10 HSPs that are included in the Somascan panel: Heat shock protein HSP 90-alpha/beta (HSP90AA1/AB1); Heat shock cognate 71 kDa protein (HSP70; HSPA8); Heat shock protein beta-1 (HSP27, HSPB1); MAP kinase-activated protein kinase 5 (MAPKAPK5); DnaJ homolog subfamily B member 1 (DNAJB1, HSP40); Stress-induced-phosphoprotein 1 (STIP1); E3 ubiquitin-protein ligase CHIP (STUB1); Ubiquitin-conjugating enzyme E2 L3 (UBE2L3); Ubiquitin-conjugating enzyme E2 N (UBE2N); and Ubiquitin-conjugating enzyme E2 G2 (UBE2G2). As such, 7 OS-related proteins and 10 HSPs are presented herein.

Statistics

Using nQuery version 9 (Statsol, Boston, MA, USA), a power analysis was performed on lipid peroxidation based upon prior reports on PCOS [50]; lipid peroxidation is the oxidative degradation of lipids to give a general measure of oxidative stress [50]. For 90% power and an alpha of 0.05 with a common standard deviation of 0.38, together with an effect size of 1.16, the required number of subjects was determined to be 17 per arm; however, this study was exploratory, as it was not possible to calculate the power for an individual OS or HSP protein, so 48 subjects were recruited to ensure adequate subject numbers. Both visual and statistical evaluations were performed on the data to assess normality. Independent t-tests were utilized when the data were normally distributed; alternatively, when the data were not normally distributed, as determined using the Kolmogorov–Smirnov test, the Mann–Whitney U non-parametric test was utilized. Bonferroni adjustment for multiple comparisons was undertaken. Correlation analyses between the vitamin D and the OS-related and HSP proteins were performed with Pearson coefficient. All the analyses were performed using Graphpad Prism version 10.0.0 (San Diego, CA, USA).

3. Results

3.1. Demographic Data

The non-obese PCOS cohort and the control women cohort were well matched for age, BMI, insulin resistance (HOMA), and inflammation (CRP), as these parameters did not differ between the two groups (Table 1). As expected, the PCOS cohort showed a classical elevation in the free androgen index and an elevated anti-Mullerian hormone. Neither 25(OH)D_3 nor 1,25$(OH)_2D_3$ differed between the groups. Of the 48 women recruited, 28 were 25(OH)D_3 deficient (<20 ng/mL (50 nmol/L) 25(OH)D_3).

3.2. Oxidative Stress Results

The results of the circulatory OS proteins are shown in Table 2 for both the non-obese PCOS and control subjects. In these non-obese, non-insulin resistant PCOS subjects, only ESD was elevated ($p < 0.01$) compared to the controls, but there were no other differences in the other OS proteins following the Bonferroni correction for multiple comparisons.

Table 2. Levels of circulatory heat shock and oxidative stress proteins in control women (n = 24) and women with polycystic ovary syndrome (PCOS) (n = 24). Levels are reported as mean ± 1 standard deviation of RFU (relative fluorescent units).

		Control	PCOS	p-Value
Heat shock proteins	HSP90AA1/AB1	6566 (11,881)	5160 (6109)	0.57
	HSP70 (HSPA8)	2140 (2361)	1606 (348)	0.23
	HSP27 (HSPB1)	2350 (1727)	3773 (4633)	0.13
	MAPKAPK5	985 (2314)	625 (504)	0.42
	DNAJB1 (HSP40)	447 (296)	485 (337)	0.66
	STIP1	7563 (7799)	6785 (4837)	0.65
	STUB1	1660 (3163)	1212 (1138)	0.48
	UBE2L3	1212 (635)	1322 (764)	0.55
	UBE2N	7696 (8053)	7014 (6284)	0.72
	UBE2G2	4642 (1853)	4401 (1002)	0.54
Oxidative stress proteins	SOD1	973 (621)	914 (325)	0.64
	GSTP1	3358 (2260)	3594 (763)	0.65
	GSTA3	209 (214)	182 (119)	0.53
	ESD	4606 (1887)	6184 (2632)	0.01
	CAT	22,445 (12,953)	22,749 (8629)	0.89
	SOD2	48,222 (10,590)	51,647 (12,763)	0.26
	SOD3	506 (455)	625 (710)	0.46

HSP90AA1/AB1: Heat shock protein HSP 90-alpha/beta; HSP70 (HSPA8): Heat shock cognate 71 kDa protein; HSP27 (HSPB1): Heat shock protein beta-1; MAPKAPK5: MAP kinase-activated protein kinase 5; DNAJB1 (HSP40): DnaJ homolog subfamily B member 1; STIP1: Stress-induced-phosphoprotein 1; STUB1: E3 ubiquitin-protein ligase CHIP; UBE2L3: Ubiquitin-conjugating enzyme E2 L3; UBE2N: Ubiquitin-conjugating enzyme E2 N; UBE2G2: Ubiquitin-conjugating enzyme E2 G2; SOD1: Superoxide dismutase; GSTP1: Glutathione S-transferase P; GSTA3: Glutathione S-transferase A3; ESD: Esterase D; CAT: Catalase; SOD2: Superoxide dismutase mitochondrial; and SOD3: Extracellular superoxide dismutase.

As neither 25(OH)D_3 nor 1,25$(OH)_2D_3$ differed between the groups, the groups were combined to increase the power of identifying an association between 25(OH)D_3 or 1,25$(OH)_2D_3$ and the OS markers. These data are shown in Table 3 and Supplemental Figures S1 and S2 for OS with 25(OH)D_3 and 1,25$(OH)_2D_3$, respectively. No associations between 25(OH)D_3 or 1,25$(OH)_2D_3$ and any of the OS proteins were found.

Table 3. Lack of correlation of circulatory heat shock and oxidative stress proteins with vitamin D_3 (25(OH)D_3) and its active form, 1,25(OH)$_2D_3$.

		25(OH)D_3		1,25(OH)$_2D_3$	
		r-Value	*p*-Value	r-Value	*p*-Value
Heat shock proteins	HSP90AA1/AB1	0.2	0.17	0.02	0.9
	HSP70 (HSPA8)	0.06	0.67	0.02	0.93
	HSP27 (HSPB1)	0.22	0.14	0.09	0.59
	MAPKAPK5	0.08	0.6	0.26	0.12
	DNAJB1 (HSP40)	0.24	0.1	0.0005	0.98
	STIP1	0.06	0.69	0.03	0.87
	STUB1	0.05	0.73	0.06	0.73
	UBE2L3	0.0002	0.96	0.03	0.91
	UBE2N	0.09	0.53	0.03	0.85
	UBE2G2	0.06	0.71	0.16	0.34
Oxidative stress proteins	SOD1	0.05	0.74	0.09	0.61
	GSTP1	0.11	0.44	0.11	0.52
	GSTA3	0.12	0.43	0.11	0.53
	ESD	0.06	0.69	0.14	0.39
	CAT	0.25	0.09	0.0003	0.99
	SOD2	0.11	0.48	0.08	0.66
	SOD3	0.17	0.25	0.21	0.22

HSP90AA1/AB1: Heat shock protein HSP 90-alpha/beta; HSP70 (HSPA8): Heat shock cognate 71 kDa protein; HSP27 (HSPB1): Heat shock protein beta-1; MAPKAPK5: MAP kinase-activated protein kinase 5; DNAJB1 (HSP40): DnaJ homolog subfamily B member 1; STIP1: Stress-induced-phosphoprotein 1; STUB1: E3 ubiquitin-protein ligase CHIP; UBE2L3: Ubiquitin-conjugating enzyme E2 L3; UBE2N: Ubiquitin-conjugating enzyme E2 N; UBE2G2: Ubiquitin-conjugating enzyme E2 G2; SOD1: Superoxide dismutase; GSTP1: Glutathione S-transferase P; GSTA3: Glutathione S-transferase A3; ESD: Esterase D; CAT: Catalase; SOD2: Superoxide dismutase mitochondrial; and SOD3: Extracellular superoxide dismutase.

When the patients were divided based on vitamin D deficiency, there were no differences in any of the OS proteins.

3.3. Heat Shock Protein Results

The results of the HSP proteins are shown in Table 2 for both the non-obese PCOS and control subjects. There were no differences in the HSPs following the Bonferroni correction for multiple comparisons in these non-obese, non-insulin resistant PCOS subjects.

As neither 25(OH)D_3 nor 1,25(OH)$_2D_3$ differed between the groups, the groups were combined to increase the power of identifying an association between 25(OH)D_3 or 1,25(OH)$_2D_3$ and the HSP markers. These data are shown in Table 3 and Supplemental Figures S3 and S4 for the HSPs with 25(OH)D_3 and 1,25(OH)$_2D_3$, respectively. There were no associations between 25(OH)D_3 or 1,25(OH)$_2D_3$ and any of the HSPs.

When the patients were divided based on vitamin D deficiency, there were no differences in any of the HSP proteins.

4. Discussion

These data show that, in non-obese PCOS subjects in which BMI systemic inflammation and insulin resistance were accounted for and did not differ to the controls, the OS proteins other than ESD and the HSPs were not elevated in PCOS. In addition, the OS proteins and the HSPs studied showed no correlation with either vitamin D, 25(OH)D_3, or its active 1,25(OH)$_2D_3$.

Increased OS is recognized as being important in the pathogenesis and pathophysiological processes of PCOS [5,6]. Esterase D (ESD) is a nonspecific esterase that detoxifies formaldehyde. Several reports have detailed that increased ESD activity is associated with various physiological and pathological processes, including autophagy, but its signaling processes are poorly understood [51]. Whilst being recognized as being dysregulated in cancer and important in inhibiting cell growth [52], ESD measurement as an OS protein in PCOS has not been described before, and here it was significantly elevated, suggesting that its function in PCOS requires further investigation. However, as the other markers of OS were unaltered, this suggests that its elevation was reflecting another (as of yet unknown) process. Malondialdehyde is a measure of OS that is derived from lipid peroxidation, has been shown to be increased in several studies, and has, additionally, been suggested to be independent of obesity [53]; the study by Kuşçu and Var is the converse to what is shown here, though their study population was not non-obese. Glutathione transferase has been reported to be low in PCOS compared to controls [54], but increased in other reports [55], whilst in this study, it did not differ. SOD removes superoxide anions and has been shown to be increased in PCOS [56], though it was unaltered in this study. This suggests that when obesity, insulin resistance, and inflammation are accounted for, this population is at no greater risk than the control population for cardiovascular risk, and that hyperandrogenemia alone is not a contributory factor to the increased oxidative stress reported in PCOS.

No differences were seen in the HSPs in this population, which is perhaps unsurprising, as their regulation would be dependent on increased OS, systemic inflammation, insulin resistance, and obesity, which were not present in this study population. The finding of elevated ESD also suggests that it is not acting as an OS protein in this context, as there was no association with any of the HSPs.

In this cohort of non-obese, non-insulin resistant PCOS patients without systemic inflammation, our findings contrast with the reported changes in the OS and HSPs in PCOS and suggest that those reported changes are due to the underlying obesity, insulin resistance, and inflammation, rather than due to the underlying pathophysiology of PCOS. Oxidative stress is associated with both obesity and insulin resistance [57,58], which may act synergistically or in an additive fashion, as an increasing BMI results in increasing insulin resistance [59]. The cardiovascular risk factors of obesity, insulin resistance, and chronic inflammation are more linked to obesity rather than the pathophysiology of PCOS; however, teasing out what the contribution of the inherent PCOS pathophysiology is versus the relative effects of insulin resistance and obesity [15,60] is only possible with populations matched to exclude confounders, including ethnicity, that may affect PCOS [61]; the contributions of obesity, insulin resistance, and inflammation were specifically accounted for to circumvent these confounders.

No correlations were seen for either the $25(OH)D_3$ or the active $1,25(OH)_2D_3$ levels with either the OS proteins or the HSPs. In addition, when the cohort was divided into vitamin D deficient and sufficient, there were no differences between the groups for either the OS or the HSPs in this cohort of non-obese individuals without insulin resistance or raised inflammatory markers. However, these data do not negate that the changes in the OS factors reported in obese PCOS are dependent on vitamin D levels or, conversely, that the deleterious effects of vitamin D deficiency are not mediated through the OS process in PCOS, as these may be affected by the obesity, inflammation, and insulin resistance that are commonly found in PCOS. A meta-analysis reported that vitamin D supplementation improved the OS parameters in PCOS [62], suggesting that vitamin D may be beneficial when OS is present. Whilst the kidneys produce 1,25-dihydroxy vitamin D for regulating the calcium and bone metabolism [63], it is recognized that $1,25(OH)_2D$ can be produced externally to the kidneys though macrophage production [64]. Consequently, 1,25-dihydroxyvitamin D blood levels may not reflect the tissue level function that could be beneficial due to the local production of 1,25-dihydroxy vitamin D [65].

This study has a number of limitations. OS proteins were measured rather than the functional serum oxidative capacity, such as measuring malondialdehyde (MDA), 4-hydroxy-nonenal (HNE), and the F2-isoprostane 15(S)-8-iso-prostaglandin F2α (15(S)-8-iso-PGF2α). However, it is recognized that these methods are difficult and less precise due to analytical issues [66]. There are no studies that could have been utilized for a power analysis for this non-obese PCOS cohort; therefore, the potential differences were based on those of an obese PCOS population; therefore, a type 2 (beta) error cannot be excluded. All the study subjects were Caucasian, and therefore these results may not be generalizable to other ethnic populations. This study lays the foundation for dissecting the initiation of OS in PCOS, as we have shown here that OS in this population does not occur; the identification of PCOS cohorts with inflammation alone and insulin resistance alone without obesity would identify the causative mechanisms for OS in this population and those scenarios where vitamin D may afford protection.

In conclusion, in a PCOS population that was non-obese and without insulin resistance and systemic inflammation, only ESD was elevated in PCOS, whilst the other OS proteins and HSPs were not elevated. Further, none of the OS proteins or HSPs were correlated with either 25(OH)D_3 or 1,25$(OH)_2D_3$ in either cohort of women, or when both cohorts were combined, indicating that the OS and HSP responses were largely absent and not affected by vitamin D in this non-obese PCOS population.

Supplementary Materials: The following supporting information can be downloaded at: https://www.mdpi.com/article/10.3390/biomedicines11072044/s1.

Author Contributions: M.N. and A.E.B. analyzed the data and wrote the manuscript. T.S. supervised clinical studies and edited the manuscript. S.L.A. contributed to study design, data interpretation and the writing of the manuscript. All authors reviewed and approved the final version of the manuscript. All authors have read and agreed to the published version of the manuscript.

Funding: This research received no external funding.

Institutional Review Board Statement: All procedures performed in studies involving human participants were in accordance with the ethical standards of the Yorkshire and The Humber NRES ethical committee, UK, and with the 1964 Helsinki declaration and its later amendments or comparable ethical standards.

Informed Consent Statement: Informed consent was obtained from all subjects involved in the study.

Data Availability Statement: All the data for this study will be made available upon reasonable request to the corresponding author.

Conflicts of Interest: No authors have any conflict of interest or competing interests to declare.

References

1. Shroff, R.; Kerchner, A.; Maifeld, M.; Van Beek, E.J.R.; Jagasia, D.; Dokras, A. Young obese women with polycystic ovary syndrome have evidence of early coronary atherosclerosis. *J. Clin. Endocrinol. Metab.* **2007**, *92*, 4609–4614. [CrossRef] [PubMed]
2. Legro, R.S.; Arslanian, S.A.; Ehrmann, D.A.; Hoeger, K.M.; Murad, M.H.; Pasquali, R.; Welt, C.K. Diagnosis and treatment of polycystic ovary syndrome: An Endocrine Society clinical practice guideline. *J. Clin. Endocrinol. Metab.* **2013**, *98*, 4565–4592. [CrossRef] [PubMed]
3. Diamanti-Kandarakis, E.; Kouli, C.R.; Bergiele, A.T.; Filandra, F.A.; Tsianateli, T.C.; Spina, G.G.; Zapanti, E.D.; Bartzis, M.I. A survey of the polycystic ovary syndrome in the Greek island of Lesbos: Hormonal and metabolic profile. *J. Clin. Endocrinol. Metab.* **1999**, *84*, 4006–4011. [CrossRef] [PubMed]
4. Sathyapalan, T.; Atkin, S.L. Recent advances in cardiovascular aspects of polycystic ovary syndrome. *Eur. J. Endocrinol.* **2012**, *166*, 575–583. [CrossRef]
5. Mohammadi, M. Oxidative Stress and Polycystic Ovary Syndrome: A Brief Review. *Int. J. Prev. Med.* **2019**, *10*, 86. [CrossRef]
6. Fernández-Sánchez, A.; Madrigal-Santillán, E.; Bautista, M.; Esquivel-Soto, J.; Morales-González, Á.; Esquivel-Chirino, C.; Durante-Montiel, I.; Sánchez-Rivera, G.; Valadez-Vega, C.; Morales-González, J.A. Inflammation, oxidative stress, and obesity. *Int. J. Mol. Sci.* **2011**, *12*, 3117–3132. [CrossRef]
7. Sathyapalan, T.; Shepherd, J.; Coady, A.M.; Kilpatrick, E.S.; Atkin, S.L. Atorvastatin reduces malondialdehyde concentrations in patients with polycystic ovary syndrome. *J. Clin. Endocrinol. Metab.* **2012**, *97*, 3951–3955. [CrossRef]
8. Bannigida, D.M.; Nayak, B.S.; Vijayaraghavan, R. Insulin resistance and oxidative marker in women with PCOS. *Arch. Physiol. Biochem.* **2020**, *126*, 183–186. [CrossRef]
9. Niinuma, S.A.; Lubbad, L.; Lubbad, W.; Moin, A.S.M.; Butler, A.E. The Role of Heat Shock Proteins in the Pathogenesis of Polycystic Ovarian Syndrome: A Review of the Literature. *Int. J. Mol. Sci.* **2023**, *24*, 1838. [CrossRef]
10. Moin, A.S.M.; Sathyapalan, T.; Butler, A.E.; Atkin, S.L. Coagulation factor dysregulation in polycystic ovary syndrome is an epiphenomenon of obesity. *Clin. Endocrinol.* **2023**, *98*, 796–802. [CrossRef]
11. Lewis, R.D.; Narayanaswamy, A.K.; Farewell, D.; Rees, D.A. Complement activation in polycystic ovary syndrome occurs in the postprandial and fasted state and is influenced by obesity and insulin sensitivity. *Clin. Endocrinol.* **2021**, *94*, 74–84. [CrossRef] [PubMed]

12. Yang, S.; Li, Q.; Song, Y.; Tian, B.; Cheng, Q.; Qing, H.; Zhong, L.; Xia, W. Serum complement C3 has a stronger association with insulin resistance than high-sensitivity C-reactive protein in women with polycystic ovary syndrome. *Fertil. Steril.* **2011**, *95*, 1749–1753. [CrossRef]
13. Moin, A.S.M.; Sathyapalan, T.; Butler, A.E.; Atkin, S.L. Classical and alternate complement factor overexpression in non-obese weight matched women with polycystic ovary syndrome does not correlate with vitamin D. *Front. Endocrinol.* **2022**, *13*, 935750. [CrossRef]
14. Butler, A.E.; Moin, A.S.M.; Sathyapalan, T.; Atkin, S.L. Components of the Complement Cascade Differ in Polycystic Ovary Syndrome. *Int. J. Mol. Sci.* **2022**, *23*, 12232. [CrossRef] [PubMed]
15. Legro, R.S. Obesity and PCOS: Implications for diagnosis and treatment. In *Seminars in Reproductive Medicine*; Thieme Medical Publishers: New York, NY, USA, 2012; Volume 30, pp. 496–506.
16. Hu, L.; Ma, L.; Xia, X.; Ying, T.; Zhou, M.; Zou, S.; Yu, H.; Yin, J. Efficacy of Bariatric Surgery in the Treatment of Women with Obesity and Polycystic Ovary Syndrome. *J. Clin. Endocrinol. Metab.* **2022**, *107*, e3217–e3229. [CrossRef] [PubMed]
17. Alalami, H.; Sathyapalan, T.; Atkin, S.L. Cardiovascular profile of pharmacological agents used for the management of polycystic ovary syndrome. *Ther. Adv. Endocrinol. Metab.* **2019**, *10*, 2042018818805674. [CrossRef] [PubMed]
18. Bastard, J.-P.; Jardel, C.; Bruckert, E.; Blondy, P.; Capeau, J.; Laville, M.; Vidal, H.; Hainque, B. Elevated levels of interleukin 6 are reduced in serum and subcutaneous adipose tissue of obese women after weight loss. *J. Clin. Endocrinol. Metab.* **2000**, *85*, 3338–3342.
19. O'Rourke, R.W.; Kay, T.; Lyle, E.A.; Traxler, S.A.; Deveney, C.W.; Jobe, B.A.; Roberts, C.T.; Marks, D.; Rosenbaum, J.T. Alterations in peripheral blood lymphocyte cytokine expression in obesity. *Clin. Exp. Immunol.* **2006**, *146*, 39–46. [CrossRef]
20. Festa, A.; D'Agostino, R., Jr.; Williams, K.; Karter, A.J.; Mayer-Davis, E.J.; Tracy, R.P.; Haffner, S.M. The relation of body fat mass and distribution to markers of chronic inflammation. *Int. J. Obes. Relat. Metab. Disord.* **2001**, *25*, 1407–1415. [CrossRef]
21. Kelly, C.C.; Lyall, H.; Petrie, J.R.; Gould, G.W.; Connell, J.M.; Sattar, N. Low grade chronic inflammation in women with polycystic ovarian syndrome. *J. Clin. Endocrinol. Metab.* **2001**, *86*, 2453–2455. [CrossRef]
22. Abdalla, M.A.; Shah, N.; Deshmukh, H.; Sahebkar, A.; Östlundh, L.; Al-Rifai, R.H.; Atkin, S.L.; Sathyapalan, T. Effect of pharmacological interventions on lipid profiles and C-reactive protein in polycystic ovary syndrome: A systematic review and meta-analysis. *Clin. Endocrinol.* **2022**, *96*, 443–459. [CrossRef]
23. Ridker, P.M.; Hennekens, C.H.; Buring, J.E.; Rifai, N. C-reactive protein and other markers of inflammation in the prediction of cardiovascular disease in women. *N. Engl. J. Med.* **2000**, *342*, 836–843. [CrossRef]
24. Fauser, B.C.; Tarlatzis, B.C.; Rebar, R.W.; Legro, R.S.; Balen, A.H.; Lobo, R.; Carmina, E.; Chang, J.; Yildiz, B.O.; Laven, J.S.; et al. Consensus on women's health aspects of polycystic ovary syndrome (PCOS): The Amsterdam ESHRE/ASRM-Sponsored 3rd PCOS Consensus Workshop Group. *Fertil. Steril.* **2012**, *97*, 28–38.e25. [CrossRef]
25. Setji, T.L.; Holland, N.D.; Sanders, L.L.; Pereira, K.C.; Diehl, A.M.; Brown, A.J. Nonalcoholic steatohepatitis and nonalcoholic Fatty liver disease in young women with polycystic ovary syndrome. *J. Clin. Endocrinol. Metab.* **2006**, *91*, 1741–1747. [CrossRef] [PubMed]
26. Baptiste, C.G.; Battista, M.C.; Trottier, A.; Baillargeon, J.P. Insulin and hyperandrogenism in women with polycystic ovary syndrome. *J. Steroid Biochem. Mol. Biol.* **2010**, *122*, 42–52. [CrossRef]
27. Chantrapanichkul, P.; Indhavivadhana, S.; Wongwananuruk, T.; Techatraisak, K.; Dangrat, C.; Sa-Nga-Areekul, N. Prevalence of type 2 diabetes mellitus compared between lean and overweight/obese patients with polycystic ovarian syndrome: A 5-year follow-up study. *Arch. Gynecol. Obstet.* **2020**, *301*, 809–816. [CrossRef]
28. Osibogun, O.; Ogunmoroti, O.; Michos, E.D. Polycystic ovary syndrome and cardiometabolic risk: Opportunities for cardiovascular disease prevention. *Trends Cardiovasc. Med.* **2020**, *30*, 399–404. [CrossRef]
29. Murri, M.; Luque-Ramírez, M.; Insenser, M.; Ojeda-Ojeda, M.; Escobar-Morreale, H.F. Circulating markers of oxidative stress and polycystic ovary syndrome (PCOS): A systematic review and meta-analysis. *Hum. Reprod. Update* **2013**, *19*, 268–288. [CrossRef]
30. Kattoor, A.J.; Pothineni, N.V.K.; Palagiri, D.; Mehta, J.L. Oxidative Stress in Atherosclerosis. *Curr. Atheroscler. Rep.* **2017**, *19*, 4[illegible] [CrossRef]
31. Liu, J.F.; Chen, P.C.; Ling, T.Y.; Hou, C.H. Hyperthermia increases HSP production in human PDMCs by stimulating ROS formation, p38 MAPK and Akt signaling, and increasing HSF1 activity. *Stem Cell Res. Ther.* **2022**, *13*, 236. [CrossRef]
32. Tsan, M.F.; Gao, B. Heat shock protein and innate immunity. *Cell. Mol. Immunol.* **2004**, *1*, 274–279. [PubMed]
33. Gao, H.; Meng, J.; Xu, M.; Zhang, S.; Ghose, B.; Liu, J.; Yao, P.; Yan, H.; Wang, D.; Liu, L. Serum Heat Shock Protein 7[illegible] Concentration in Relation to Polycystic Ovary Syndrome in a Non-Obese Chinese Population. *PLoS ONE* **2013**, *8*, e6772[illegible] [CrossRef] [PubMed]
34. Costes, S.; Huang, C.-J.; Gurlo, T.; Daval, M.; Matveyenko, A.V.; Rizza, R.A.; Butler, A.E.; Butler, P.C. β-cell dysfunctional ERAD/ubiquitin/proteasome system in type 2 diabetes mediated by islet amyloid polypeptide-induced UCH-L1 deficiency. *Diabetes* **2011**, *60*, 227–238. [CrossRef] [PubMed]
35. Vigouroux, S.; Briand, M.; Briand, Y. Linkage between the proteasome pathway and neurodegenerative diseases and aging. *Mol. Neurobiol.* **2004**, *30*, 201–221. [CrossRef]
36. Hooper, P.L.; Hooper, P.L. Inflammation, heat shock proteins, and type 2 diabetes. *Cell Stress Chaperones* **2009**, *14*, 113–11[illegible] [CrossRef]

37. Qi, D.; Liu, H.; Niu, J.; Fan, X.; Wen, X.; Du, Y.; Mou, J.; Pei, D.; Liu, Z.; Zong, Z.; et al. Heat shock protein 72 inhibits c-Jun N-terminal kinase 3 signaling pathway via Akt1 during cerebral ischemia. *J. Neurol. Sci.* **2012**, *317*, 123–129. [CrossRef]
38. Hahn, S.; Haselhorst, U.; Tan, S.; Quadbeck, B.; Schmidt, M.; Roesler, S.; Kimmig, R.; Mann, K.; Janssen, O. Low serum 25-hydroxyvitamin D concentrations are associated with insulin resistance and obesity in women with polycystic ovary syndrome. *Exp. Clin. Endocrinol. Diabetes* **2006**, *114*, 577–583. [CrossRef]
39. Li, H.W.; Brereton, R.E.; Anderson, R.A.; Wallace, A.M.; Ho, C.K. Vitamin D deficiency is common and associated with metabolic risk factors in patients with polycystic ovary syndrome. *Metabolism Clin. Exp.* **2011**, *60*, 1475–1481. [CrossRef]
40. Butler, A.E.; Dargham, S.R.; Abouseif, A.; El Shewehy, A.; Atkin, S.L. Vitamin D deficiency effects on cardiovascular parameters in women with polycystic ovary syndrome: A retrospective, cross-sectional study. *J. Steroid Biochem. Mol. Biol.* **2021**, *211*, 105892. [CrossRef]
41. Lerchbaum, E.; Obermayer-Pietsch, B. Vitamin D and fertility: A systematic review. *Eur. J. Endocrinol./Eur. Fed. Endocr. Soc.* **2012**, *166*, 765–778. [CrossRef]
42. Selimoglu, H.; Duran, C.; Kiyici, S.; Ersoy, C.; Guclu, M.; Ozkaya, G.; Tuncel, E.; Erturk, E.; Imamoglu, S. The effect of vitamin D replacement therapy on insulin resistance and androgen levels in women with polycystic ovary syndrome. *J. Endocrinol. Investig.* **2010**, *33*, 234–238. [CrossRef] [PubMed]
43. Krul-Poel, Y.H.M.; Snackey, C.; Louwers, Y.; Lips, P.T.A.M.; Lambalk, C.B.; Laven, J.S.E.; Simsek, S. The role of vitamin D in metabolic disturbances in polycystic ovary syndrome (PCOS): A systematic review. *Eur. J. Endocrinol./Eur. Fed. Endocr. Soc.* **2013**, *169*, 853–865. [CrossRef] [PubMed]
44. Wimalawansa, S.J. Vitamin D Deficiency: Effects on Oxidative Stress, Epigenetics, Gene Regulation, and Aging. *Biology* **2019**, *8*, 30. [CrossRef] [PubMed]
45. Cunningham, T.K.; Allgar, V.; Dargham, S.R.; Kilpatrick, E.; Sathyapalan, T.; Maguiness, S.; Mokhtar Rudin, H.R.; Abdul Ghani, N.M.; Latiff, A.; Atkin, S.L. Association of Vitamin D Metabolites with Embryo Development and Fertilization in Women with and without PCOS Undergoing Subfertility Treatment. *Front. Endocrinol.* **2019**, *10*, 13. [CrossRef] [PubMed]
46. Rotterdam ESHRE/ASRM-Sponsored PCOS Consensus Workshop Group. Revised 2003 consensus on diagnostic criteria and long-term health risks related to polycystic ovary syndrome (PCOS). *Human Reprod.* **2004**, *19*, 41–47. [CrossRef]
47. Webb, A.R.; Kazantzidis, A.; Kift, R.C.; Farrar, M.D.; Wilkinson, J.; Rhodes, L.E. Meeting Vitamin D Requirements in White Caucasians at UK Latitudes: Providing a Choice. *Nutrients* **2018**, *10*, 497. [CrossRef]
48. Kahal, H.; Halama, A.; Aburima, A.; Bhagwat, A.M.; Butler, A.E.; Graumann, J.; Suhre, K.; Sathyapalan, T.; Atkin, S.L. Effect of induced hypoglycemia on inflammation and oxidative stress in type 2 diabetes and control subjects. *Sci. Rep.* **2020**, *10*, 4750. [CrossRef]
49. Kraemer, S.; Vaught, J.D.; Bock, C.; Gold, L.; Katilius, E.; Keeney, T.R.; Kim, N.; Saccomano, N.A.; Wilcox, S.K.; Zichi, D.; et al. From SOMAmer-based biomarker discovery to diagnostic and clinical applications: A SOMAmer-based, streamlined multiplex proteomic assay. *PLoS ONE* **2011**, *6*, e26332. [CrossRef]
50. Torun, A.N.; Vural, M.; Cece, H.; Camuzcuoglu, H.; Toy, H.; Aksoy, N. Paraoxonase-1 is not affected in polycystic ovary syndrome without metabolic syndrome and insulin resistance, but oxidative stress is altered. *Gynecol. Endocrinol.* **2011**, *27*, 988–992. [CrossRef]
51. Yang, Y.; Chen, X.; Yao, W.; Cui, X.; Li, N.; Lin, Z.; Zhao, B.; Miao, J. Esterase D stabilizes FKBP25 to suppress mTORC1. *Cell. Mol. Biol. Lett.* **2021**, *26*, 50. [CrossRef]
52. Yao, W.; Yang, Y.; Chen, X.; Cui, X.; Zhou, B.; Zhao, B.; Lin, Z.; Miao, J. Activation of Esterase D by FPD5 Inhibits Growth of A549 Lung Cancer Cells via JAB1/p53 Pathway. *Genes* **2022**, *13*, 786. [CrossRef] [PubMed]
53. Kuşçu, N.K.; Var, A. Oxidative stress but not endothelial dysfunction exists in non-obese, young group of patients with polycystic ovary syndrome. *Acta Obstet. Gynecol. Scand.* **2009**, *88*, 612–617. [CrossRef] [PubMed]
54. Uçkan, K.; Demir, H.; Başkiran, Y.; Demir, C. Investigation of activities enzyme prolidase (pro) and glutathione s-transferase (gst) in polycystic ovary syndrome (pcos) patients. *J. Sci. Rep. A* **2022**, *050*, 20–31.
55. Fatima, Q.; Amin, S.; Kawa, I.A.; Jeelani, H.; Manzoor, S.; Rizvi, S.M.; Rashid, F. Evaluation of antioxidant defense markers in relation to hormonal and insulin parameters in women with polycystic ovary syndrome (PCOS): A case-control study. *Diabetes Metab. Syndr.* **2019**, *13*, 1957–1961. [CrossRef] [PubMed]
56. Sabuncu, T.; Vural, H.; Harma, M.; Harma, M. Oxidative stress in polycystic ovary syndrome and its contribution to the risk of cardiovascular disease. *Clin. Biochem.* **2001**, *34*, 407–413. [CrossRef] [PubMed]
57. Pérez-Torres, I.; Castrejón-Téllez, V.; Soto, M.E.; Rubio-Ruiz, M.E.; Manzano-Pech, L.; Guarner-Lans, V. Oxidative Stress, Plant Natural Antioxidants, and Obesity. *Int. J. Mol. Sci.* **2021**, *22*, 1786. [CrossRef]
58. Yazıcı, D.; Sezer, H. Insulin Resistance, Obesity and Lipotoxicity. *Adv. Exp. Med. Biol.* **2017**, *960*, 277–304.
59. Gobato, A.O.; Vasques, A.C.; Zambon, M.P.; Barros Filho Ade, A.; Hessel, G. Metabolic syndrome and insulin resistance in obese adolescents. *Rev. Paul. Pediatr.* **2014**, *32*, 55–62. [CrossRef]
60. Moin, A.S.M.; Sathyapalan, T.; Diboun, I.; Elrayess, M.A.; Butler, A.E.; Atkin, S.L. Metabolic consequences of obesity on the hypercoagulable state of polycystic ovary syndrome. *Sci. Rep.* **2021**, *11*, 5320. [CrossRef]
61. Butler, A.E.; Abouseif, A.; Dargham, S.R.; Sathyapalan, T.; Atkin, S.L. Metabolic comparison of polycystic ovarian syndrome and control women in Middle Eastern and UK Caucasian populations. *Sci. Rep.* **2020**, *10*, 18895. [CrossRef]

62. Akbari, M.; Ostadmohammadi, V.; Lankarani, K.B.; Tabrizi, R.; Kolahdooz, F.; Heydari, S.T.; Kavari, S.H.; Mirhosseini, N.; Mafi, A.; Dastorani, M.; et al. The Effects of Vitamin D Supplementation on Biomarkers of Inflammation and Oxidative Stress among Women with Polycystic Ovary Syndrome: A Systematic Review and Meta-Analysis of Randomized Controlled Trials. *Horm. Metab. Res.* **2018**, *50*, 271–279. [CrossRef] [PubMed]
63. Adams, J.S.; Ramin, J.; Rafison, B.; Windon, C.; Windon, A.; Liu, P.T. Redefining Human Vitamin D Sufficiency: Back to the Basics. *Bone Res.* **2013**, *1*, 2–10. [CrossRef] [PubMed]
64. Adams, J.S.; Rafison, B.; Witzel, S.; Reyes, R.E.; Shieh, A.; Chun, R.; Zavala, K.; Hewison, M.; Liu, P.T. Regulation of the extrarenal CYP27B1-hydroxylase. *J. Steroid Biochem. Mol. Biol.* **2014**, *144 Pt A*, 22–27. [CrossRef]
65. Adams, I.P.; Dack, S.; Dickinson, F.M.; Ratledge, C. The distinctiveness of ATP:citrate lyase from Aspergillus nidulans. *Biochim Biophys. Acta* **2002**, *1597*, 36–41. [CrossRef]
66. Tsikas, D. Assessment of lipid peroxidation by measuring malondialdehyde (MDA) and relatives in biological samples: Analytical and biological challenges. *Anal. Biochem.* **2017**, *524*, 13–30. [CrossRef] [PubMed]

MDPI AG
Grosspeteranlage 5
4052 Basel
Switzerland
Tel.: +41 61 683 77 34

Biomedicines Editorial Office
E-mail: biomedicines@mdpi.com
www.mdpi.com/journal/biomedicines

www.ingramcontent.com/pod-product-compliance
Lightning Source LLC
Chambersburg PA
CBHW062057220926
51850CB00010B/2793
9783725856756